工业和信息化精品系列教材
云计算技术

Cloud Computing Technology

微课版

OpenStack
云计算管理平台
项目教程

崔升广 ● 主编
周晓姝 于淼 杨宇 单立娟 ● 副主编

人民邮电出版社
北京

图书在版编目（CIP）数据

OpenStack云计算管理平台项目教程：微课版／崔升广主编． -- 北京：人民邮电出版社，2021.8（2024.6重印）
工业和信息化精品系列教材．云计算技术
ISBN 978-7-115-56637-9

Ⅰ．①O… Ⅱ．①崔… Ⅲ．①云计算－高等学校－教材 Ⅳ．①TP393.027

中国版本图书馆CIP数据核字(2021)第111490号

内 容 提 要

本书基于云计算应用实际需求，由浅入深、全面系统地讲解了 OpenStack 云计算管理平台的相关内容。本书共有 8 个项目，包括 OpenStack 云计算基础、OpenStack 安装与部署、OpenStack 认证服务、OpenStack 镜像服务、OpenStack 网络服务、OpenStack 计算服务、OpenStack 存储服务，以及 OpenStack 高级控制服务。本书内容丰富，注重系统性、实用性和可操作性，每个知识点都有相应的操作示例，便于读者快速掌握。

本书可作为高校计算机相关专业的教材，也可作为广大计算机爱好者自学 OpenStack 云计算管理平台的参考用书，还可作为云计算运维与管理的参考用书及社会培训教材。

◆ 主　　编　崔升广
　 副 主 编　周晓姝　于　淼　杨　宇　单立娟
　 责任编辑　郭　雯
　 责任印制　王　郁　彭志环
◆ 人民邮电出版社出版发行　　北京市丰台区成寿寺路 11 号
　 邮编 100164　电子邮件 315@ptpress.com.cn
　 网址 https://www.ptpress.com.cn
　 北京天宇星印刷厂印刷
◆ 开本：787×1092　1/16
　 印张：17.25　　　　　　　　　　2021 年 8 月第 1 版
　 字数：476 千字　　　　　　　　　2024 年 6 月北京第 9 次印刷

定价：59.80 元

读者服务热线：(010)81055256　印装质量热线：(010)81055316
反盗版热线：(010)81055315
广告经营许可证：京东市监广登字 20170147 号

前言 FOREWORD

近年来，互联网产业飞速发展，云计算作为一种弹性计算机资源服务的提供方式应运而生。云计算提供的计算机资源服务是与水、电、天然气等类似的公共资源服务。目前，通过技术发展和经验积累，云计算技术和产业已进入一个相对成熟的阶段，成为当前信息技术产业发展和创新的热点。

现在已经有许多云计算管理平台面向公众提供云计算服务，相关行业迫切需要云计算技术人才，特别是需要熟练掌握云计算管理平台规划、部署和运维的高端应用型人才。当前许多高校，特别是高职院校已经开设或正在筹备开设云计算专业，多数院校已经将云计算或大数据的课程纳入传统的计算机网络技术、通信技术、计算机应用技术等专业的人才培养方案。

OpenStack 特别适合用来开展云计算的教学和实验工作。我国很多高等院校将"云计算应用技术"作为一门重要的专业课程，编者编写本书的目的是全面、系统地讲授这门课程，使学生能够熟悉云计算的原理，掌握云计算管理平台的部署和运维的方法及技能。

本书融入了编者丰富的教学经验和多位长期从事云计算运维工作的资深工程师的实践经验，从云计算初学者的视角出发，采用"教、学、做一体化"的教学方法，为培养高端应用型人才提供合适的教学与训练教材。本书以实际项目转化的案例为主线，以"学做合一"的理念为指导，在完成技术讲解的同时，对读者提出相应的自学要求并提供指导。在学习本书的过程中，读者不仅能够完成快速入门的基本技术学习，还能够进行实际项目的开发与实现。

本书的主要特点如下。

（1）内容丰富、技术新颖，图文并茂、通俗易懂，具有很强的实用性。

（2）组织合理、有效。按照由浅入深的顺序，在逐渐丰富系统功能的同时，引入相关技术与知识，实现了技术讲解与训练的合二为一，有助于"教、学、做一体化"教学方法的实施。

本书的内容紧紧围绕着实际项目进行，为了使读者快速地掌握相关技术并按实际项目开发要求熟练运用，本书在各章重要知识点后面都根据实际项目设计了相关实例配置，介绍了要实现的项目功能，以完成详细配置。

为方便读者使用，书中全部实例的源代码及电子教案均免费赠送，读者可登录人民邮电出版社的人邮教育社区（www.ryjiaoyu.com）进行下载。

本书由崔升广任主编，周晓姝、于淼、杨宇、单立娟任副主编，周晓姝、于淼、杨宇、单立娟编写了项目 1 和项目 2，崔升广编写了项目 3～项目 8，崔升广负责全书的统稿和定稿。

由于编者水平有限,书中疏漏和不足之处在所难免,殷切希望广大读者批评指正,编者将不胜感激,编者电子邮箱为 84813752@qq.com。

编 者

2021 年 3 月

目录 CONTENTS

项目 1

OpenStack 云计算基础 ……………………………………………………… 1

- 1.1 项目陈述 ………………………………………………………………………… 1
- 1.2 必备知识 ………………………………………………………………………… 1
  - 1.2.1 云计算概述 …………………………………………………………… 1
  - 1.2.2 虚拟化技术 …………………………………………………………… 5
  - 1.2.3 OpenStack 概述 ……………………………………………………… 13
- 1.3 项目实施 ………………………………………………………………………… 18
  - 1.3.1 VMware Workstation 安装 ………………………………………… 18
  - 1.3.2 虚拟机安装 …………………………………………………………… 20
- 课后习题 ……………………………………………………………………………… 26

项目 2

OpenStack 安装与部署 ……………………………………………………… 27

- 2.1 项目陈述 ………………………………………………………………………… 27
- 2.2 必备知识 ………………………………………………………………………… 27
  - 2.2.1 Linux 相关知识 ……………………………………………………… 27
  - 2.2.2 云计算管理平台部署需求与规划 …………………………………… 43
- 2.3 项目实施 ………………………………………………………………………… 45
  - 2.3.1 使用 Packstack 一键部署 OpenStack 云计算管理平台 ………… 45
  - 2.3.2 通过 Dashboard 体验 OpenStack 云计算管理平台功能 ………… 54
- 课后习题 ……………………………………………………………………………… 56

项目 3

OpenStack 认证服务 ………………………………………………………… 57

- 3.1 项目陈述 ………………………………………………………………………… 57
- 3.2 必备知识 ………………………………………………………………………… 57
  - 3.2.1 认证服务基础 ………………………………………………………… 57
  - 3.2.2 认证服务身份管理 …………………………………………………… 61
- 3.3 项目实施 ………………………………………………………………………… 65
  - 3.3.1 基于 Dashboard 界面管理项目、用户、组和角色 ……………… 65
  - 3.3.2 基于命令行界面管理项目、用户、组和角色 …………………… 77
- 课后习题 ……………………………………………………………………………… 88

项目 4

OpenStack 镜像服务 ………………………………………………… 89

- 4.1 项目陈述 …………………………………………………………………… 89
- 4.2 必备知识 …………………………………………………………………… 89
  - 4.2.1 镜像服务基础 ……………………………………………………… 89
  - 4.2.2 镜像、实例与镜像元数据 ………………………………………… 95
- 4.3 项目实施 …………………………………………………………………… 99
  - 4.3.1 基于 Web 界面管理镜像服务 …………………………………… 99
  - 4.3.2 基于命令行界面管理镜像服务 ………………………………… 102
- 课后习题 ………………………………………………………………………… 112

项目 5

OpenStack 网络服务 ……………………………………………… 114

- 5.1 项目陈述 ………………………………………………………………… 114
- 5.2 必备知识 ………………………………………………………………… 114
  - 5.2.1 网络虚拟化 ……………………………………………………… 114
  - 5.2.2 OpenStack 网络服务基础 ……………………………………… 118
  - 5.2.3 Neutron 主要插件、代理与服务 ………………………………… 122
- 5.3 项目实施 ………………………………………………………………… 127
  - 5.3.1 基于 Web 界面管理网络服务 …………………………………… 127
  - 5.3.2 基于命令行界面管理网络服务 ………………………………… 137
- 课后习题 ………………………………………………………………………… 148

项目 6

OpenStack 计算服务 ……………………………………………… 150

- 6.1 项目陈述 ………………………………………………………………… 150
- 6.2 必备知识 ………………………………………………………………… 150
  - 6.2.1 OpenStack 计算服务基础 ……………………………………… 150
  - 6.2.2 Nova 部署架构 …………………………………………………… 157
  - 6.2.3 Nova 的元数据工作机制 ………………………………………… 161
  - 6.2.4 虚拟机实例管理 ………………………………………………… 164
- 6.3 项目实施 ………………………………………………………………… 166
  - 6.3.1 基于 Web 界面管理计算服务 …………………………………… 166
  - 6.3.2 基于命令行界面管理计算服务 ………………………………… 176
- 课后习题 ………………………………………………………………………… 183

项目 7

OpenStack 存储服务 ·················· 185

- 7.1 项目陈述 ·················· 185
- 7.2 必备知识 ·················· 185
  - 7.2.1 Cinder 块存储服务基础 ·················· 185
  - 7.2.2 Swift 对象存储服务基础 ·················· 192
- 7.3 项目实施 ·················· 206
  - 7.3.1 基于 Web 界面管理存储服务 ·················· 206
  - 7.3.2 基于命令行界面管理存储服务 ·················· 225
- 课后习题 ·················· 241

项目 8

OpenStack 高级控制服务 ·················· 244

- 8.1 项目陈述 ·················· 244
- 8.2 必备知识 ·················· 245
  - 8.2.1 Telemetry 计量与监控服务基础 ·················· 245
  - 8.2.2 Ceilometer 数据收集服务 ·················· 246
  - 8.2.3 Gnocchi 资源索引和计量存储服务 ·················· 250
  - 8.2.4 Aodh 警告服务 ·················· 252
  - 8.2.5 Heat 编排服务基础 ·················· 253
- 8.3 项目实施 ·················· 255
  - 8.3.1 基于 Web 界面管理高级控制服务 ·················· 255
  - 8.3.2 基于命令行界面管理高级控制服务 ·················· 261
- 课后习题 ·················· 268

项目 1
OpenStack 云计算基础

【学习目标】

- 了解云计算的起源以及云计算的基本概念。
- 理解虚拟化技术的基本概念以及 OpenStack 所支持的虚拟化技术。
- 了解 OpenStack 的起源及其版本演变历程。
- 理解 OpenStack 的架构。
- 掌握 VMware Workstation 以及虚拟机的安装方法。

1.1 项目陈述

OpenStack 是云操作系统,用于部署云计算管理平台,学习 OpenStack 首先需要了解云计算(Cloud Computing)的基本知识,理解相关概念与理论。2020 年 5 月 13 日,OpenStack 发布了第 21 个版本,即 Ussuri,而距 OpenStack 发布第 1 个版本 Austin 仅约 10 年,可见 OpenStack 的发展是非常迅速的,当然,它的发展离不开各大厂商的支持,也得益于当前社会经济发展的驱动。下面开始揭开 OpenStack 的神秘面纱。

1.2 必备知识

1.2.1 云计算概述

云计算提供的计算机资源服务是与水、电、天然气等类似的公共资源的服务。亚马逊云计算服务(Amazon Web Services,AWS)提供专业的云计算服务,于 2006 年推出,以 Web 服务的形式向企业提供 IT 基础设施服务,通常称为云计算,其主要优势之一是能够根据业务发展来扩展较低可变成本以替代前期资本基础设施费用,已成为公有云的事实标准。OpenStack 是开源云计算管理平台的一面旗帜,也已经成为开源云架构的事实标准。

1. 云计算的起源

1959 年,克里斯托弗·斯特雷奇(Christopher Strachey)提出了虚拟化的基本概念。2006 年 3 月,亚马逊公司首先提出弹性计算云服务。2006 年 8 月,谷歌首席执行官埃里克·施密特(Eric Schmidt)在搜索引擎大会上首次

V1-1 云计算的起源

提出了"云计算"的概念,从那时候起,云计算开始受到关注,这也标志着云计算的诞生。2010年,中华人民共和国工业和信息化部(简称工信部)、中华人民共和国发展和改革委员会联合印发了《关于做好云计算服务创新发展试点示范工作的通知(发改高技〔2010〕2480号)》,2015年,工信部发布了《工业和信息化部办公厅关于印发〈云计算综合标准化体系建设指南〉的通知》,相关文件的陆续发布,为云计算的发展奠定了基础。

云计算经历了从集中时代向网络时代、分布式时代的演变,并最终在分布式基础之上形成了云时代,如图1.1所示。

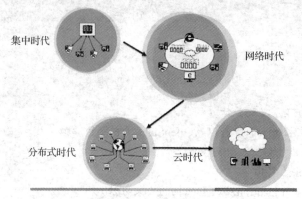

图1.1 云计算的演变

云计算作为一种计算技术和服务理念,有着极其浓厚的技术背景。谷歌公司作为搜索公司,首创云计算这一概念有着很大的必然性。随着众多互联网厂商的兴起,各家互联网公司对云计算的投入和研发不断加深,陆续形成了完整的云计算技术架构、硬件网络,而服务器方面逐步向数据中心、全球网络互连、软件系统等方向发展,操作系统、文件系统、并行计算架构、并行计算数据库和开发工具等云计算系统关键部件得到完善。

云计算的最终目标是将计算、服务和应用作为一种公共设施提供给公众,使人们能够像使用水、电、天然气等那样便捷地使用计算资源。

2. 云计算的基本概念

相信读者都听到过阿里云、华为云、百度云、腾讯云等,那么到底什么是云计算?云计算又能做什么呢?云计算是一种基于网络的超级计算模式,基于用户的不同需求提供所需要的资源,包括计算资源、网络资源、存储资源等。云计算服务通常运行在若干台高性能物理服务器之上,可提供如每秒10万亿次级的运算能力,可以用来模拟核爆炸、预测气候变化以及市场发展趋势等。

(1)云计算的定义

云计算将计算任务分布在用大量计算机构成的资源池上,使各种应用系统能够根据需要获取计算力、存储空间和各种软件服务,这种资源池称为"云"。"云"是一些可以自我维护和管理的虚拟计算资源,通常为一些大型服务器集群,包括计算服务器、存储服务器、宽带资源等,云计算将所有的计算资源集中起来,并由软件实现自动管理,无须人为参与。之所以称为"云",是因为它在某些方面具有与现实中云类似的特征:云一般较大;云可以动态伸缩,它的边界是模糊的;云在空中飘忽不定,无法也无须确定它的具体位置,但它确实存在于某处。

V1-2 云计算的定义

云计算有狭义和广义之分。

狭义上讲,"云"实质上就是一个网络,云计算就是一种提供资源的网络,包括硬件、软件和平台,使用者可以随时获取"云"上的资源,按需求量使用,并且它可以看作无限扩展的,只要按使用量付费就可以。"云"就像自来水厂一样,人们可以随时接水,且不限量,按照用水量付费给自来水厂就可以,在用户看来,水资源是无限的。

广义上讲,云计算是与信息技术、软件、互联网相关的一种服务,用户通过网络以按需、易扩展的方式获得所需要的服务。这种计算资源共享池叫作"云",云计算把许多计算资源集合起来,通过软件实现自动化管理,只需要很少的人参与,就能让资源被快速提供。也就是说,计算能力作为一种商品,可以在互联网上流通,就像水、电、天然气一样,可以被人们方便地取用,且价格较为低廉,这种服务可以是与信息技术、软件和互联网相关的服务。

总之,云计算不是一种全新的网络技术,而是一种全新的网络应用。云计算指以互联网为中心,在网站上提供快速且安全的云计算服务与数据存储服务,让每一个使用互联网的人都可以使用网络上的庞大计算资源。

云计算是继计算机、互联网之后一种新的信息技术革新,是信息时代的一个巨大飞跃。虽然目前有关云计算的定义有很多,但总体上来说,云计算的基本含义是一致的,即云计算具有很强的扩展性和需要性,可以为用户提供一种全新的体验。云计算的核心是将很多的计算机资源协调在一起,使用户通过网络就可以获取无限的资源,同时获取的资源几乎不受时间和空间的限制。

V1-3 云计算的服务模式

(2)云计算的服务模式

云计算的服务模式由3部分组成,包括基础设施即服务(Infrastructure as a Service,IaaS)、平台即服务(Platform as a Service,PaaS)和软件即服务(Software as a Service,SaaS),如图1.2所示。

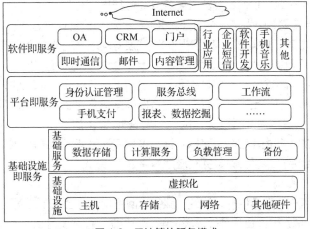

图 1.2 云计算的服务模式

① IaaS。什么是基础设施呢?服务器、硬盘、交换机等物理设备都是基础设施。云计算服务提供商购买服务器、硬盘、网络设施等来搭建基础服务,人们便可以在云计算管理平台上根据需求购买相应计算能力的内存空间、磁盘空间、网络带宽,搭建自己的云计算管理平台。这类云计算服务提供商的典型代表便是阿里云、腾讯云、华为云等。

优点:IaaS 能够根据业务需求灵活配置所需,扩展伸缩方便。

缺点:IaaS 开发维护需要投入较多人力,专业性要求较高。

② PaaS。什么是平台呢?可以将平台理解成中间件,这类云计算厂商在基础设施上进行开发,搭建操作系统,提供了一套完整的应用解决方案。开发大多数所需中间件服务,如 MySQL 数据库

服务、RocketMQ 服务等时，无须深度开发，专注业务代码即可。其典型代表便是 Pivatal Cloud Foundary、Google App Engine 等。

优点：PaaS 无须开发中间件，所需即所用，能够快速使用；部署快速，人力投入较少。

缺点：PaaS 的灵活通用性较低，过度依赖平台。

③ SaaS。软件即服务是大多数人每天都会接触到的，如办公自动化（Office Automation，OA）系统、微信公众平台。SaaS 可直接通过互联网为用户提供软件和应用程序的服务，用户可通过租赁的方式获取安装在厂商或者服务供应商上的软件。虽然这些服务用于商业或者娱乐，但是它也属于云计算的一部分，一般面向对象是普通用户，最常见的服务模式是提供给用户一组账号和密码。

优点：SaaS 所见即所得，无须开发。

缺点：SaaS 需定制，无法快速满足个性化需求。

IaaS 主要对应基础设施，可实现底层资源虚拟化以及实际云应用平台部署，这是一个网络架构由规划架构到最终物理实现的过程。PaaS 基于 IaaS 技术和平台，可部署终端用户使用的应用或程序，提供对外服务的接口或服务产品，最终实现整个平台的管理和平台的可伸缩化。SaaS 基于现成的 PaaS，作为终端用户最后接触的产品，完成现有资源的对外服务以及服务的租赁化。

（3）云计算的部署模式

云计算的部署模式通常有以下几种。

① 公有云：在这种模式下，应用程序、资源、存储和其他服务都由云服务提供商提供给用户，这些服务多半是免费的，也有部分按需或按使用量来付费，这种模式只能通过互联网来访问和使用。同时，这种模式在私人信息和数据保护方面比较有保证，通常可以提供可扩展的云服务并能进行高效设置。

V1-4　云计算的部署模式

② 私有云：这种模式专门为某一个企业服务，不管是自己管理还是第三方管理，是自己负责还是第三方负责，只要使用的方式没有问题，就能为企业带来很显著的帮助。但这种模式所要面临的问题是，纠正、检查等安全问题需企业自己负责，出了问题只能企业自己承担后果。此外，整套系统需要企业自己出钱购买、建设和管理。这种云计算部署模式可广泛地产生正面效益，从模式的名称也可看出，它可以为所有者提供具备充分优势和功能的服务。

③ 混合云：混合云是两种或两种以上的云计算部署模式的混合体，如公有云和私有云混合。它们相互独立，但在云的内部相互结合，可以发挥出多种云计算部署模式各自的优势；可使用标准的或专有的技术将它们组合起来，具有数据和应用程序的可移植性。

（4）云计算的生态系统

云计算的生态系统主要涉及硬件、软件、服务、网络和云安全 5 个方面，如图 1.3 所示。

① 硬件。云计算相关硬件包括基础环境设备、服务器、存储设备、网络设备等数据中心设备，以及提供和使用云服务的终端设备。

② 软件。云计算相关软件主要包括资源调度和管理系统、平台和应用软件等。

③ 服务。服务包括云服务和面向云计算系统建设应用的云支撑服务。

④ 网络。云计算具有泛在网络的访问特性，用户无论是通过互联网、电信网还是广播电视网，都能够使用云服务。

⑤ 云安全。云安全包括网络安全、系统安全、服务安全、应用安全等。云安全涉及服务可用性、数据机密性和完整性、隐私保护、物理安全、恶意攻击防范等诸多方面，是影响云计算发展的关键因素之一。

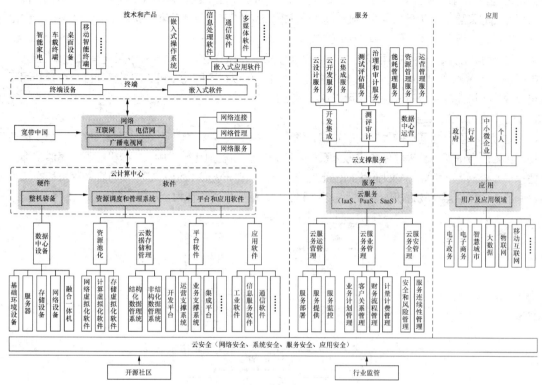

图 1.3 云计算的生态系统

1.2.2 虚拟化技术

虚拟化是指为运行的程序或软件营造它所需要的运行环境。在采用虚拟化技术后,程序或软件的运行不再独享底层的物理计算资源,它只是运行在一个虚拟化的计算资源中,而对底层的影响可能与之前所运行的计算机结构完全不同。虚拟化的主要目的是对 IT 基础设施和资源管理方式进行简化。虚拟化的消费者可以是最终用户、应用程序、操作系统、访问资源或与资源交互相关的其他服务。虚拟化是云计算的基础,虚拟化使得在一台物理服务器上可以运行多台虚拟机,虚拟机共享物理机的 CPU、内存、输入/输出(Input/Output,I/O)硬件资源,但逻辑上虚拟机之间是相互隔离的。IaaS 是基础设施架构平台,可实现底层资源虚拟化,云计算、OpenStack 都离不开虚拟化,因为虚拟化是云计算重要的支撑技术之一。OpenStack 作为 IaaS 云操作系统,最主要的服务就是为用户提供虚拟机,在目前 OpenStack 的实际应用中,主要使用 KVM 和 Xen 这两种 Linux 虚拟化技术。

1. 虚拟化的基本概念

(1)虚拟化的定义

虚拟化把物理资源转变为逻辑上可以管理的资源,以打破物理结构之间的壁垒,让资源在虚拟的而不是真实的环境中运行,是一个可简化管理、优化资源的解决方案。虚拟化让所有的资源都运行在各种各样的物理平台上,资源的管理都将按逻辑方式进行,完全实现资源的自动化分配,而虚拟化技术就是实现它的理想工具。

V1-5 虚拟化的定义

① 虚拟化前。一台主机对应一个操作系统，后台多个应用程序会对特定的资源进行争抢，存在相互冲突的风险；在实际情况中，业务系统与硬件绑定，不能灵活部署；就数据的统计来说，虚拟化前的系统资源利用率一般只有15%左右。

② 虚拟化后。一台主机可以虚拟出多个操作系统，独立的操作系统和应用拥有独立的CPU、内存和I/O资源，相互隔离；业务系统独立于硬件，可以在不同的主机之间进行迁移；充分利用系统资源，对机器的系统资源利用率可以达到60%。

（2）虚拟化体系结构

虚拟化主要是指通过软件实现的方案，常见的虚拟化体系结构如图1.4所示。这是一个直接在物理机上运行虚拟机管理程序的虚拟化系统。在x86平台虚拟化技术中，虚拟机管理程序通常称为虚拟机监控器（Virtual Machine Monitor，VMM），或Hypervisor。它是运行在物理机和虚拟机之间的一个软件层，物理机通常称为主机（Host），虚拟机通常称为客户机（Client）。

图1.4 常见的虚拟化体系结构

① 物理机。物理机指物理存在的计算机，又称宿主计算机。当虚拟机嵌套时，运行虚拟机的虚拟机也是宿主机，但不是物理机。主机操作系统是指物理机的操作系统，在主机操作系统上安装的虚拟机软件可以在计算机上模拟一台或多台虚拟机。

② 虚拟机。虚拟机指在物理机上运行的操作系统中模拟出来的计算机，又称客户机，理论上完全等同于实体的物理机。每个虚拟机都可以安装自己的操作系统或应用程序，并连接网络，运行在虚拟机上的操作系统称为客户操作系统。

Hypervisor基于主机的硬件资源给虚拟机提供了一个虚拟的操作平台并管理每个虚拟机的运行，所有虚拟机独立运行并共享主机的所有硬件资源。Hypervisor是提供虚拟机硬件模拟的专门软件，可分为两类：原生型（Native）和宿主型（Hosted）。

① 原生型。原生型又称裸机型（Bare-metal），Hypervisor作为一个精简的操作系统（操作系统也是软件，只不过它是一种比较特殊的软件），直接运行在硬件之上以控制硬件资源并管理虚拟机，比较常见的有VMware ESXi、Microsoft Hyper-V等。

② 宿主型。宿主型又称托管型，Hypervisor运行在传统的操作系统之上，同样可以模拟出一整套虚拟硬件平台，比较常见的有VMware Workstation、Oracle Virtual Box等。

从性能角度来看，无论是原生型还是宿主型都会有性能损耗，但宿主型的损耗比原生型的更大，所以企事业生产环境中基本使用的是原生型的 Hypervisor，宿主型的 Hypervisor 一般用于实验或测试环境。

（3）虚拟化分类

虚拟化分类包括平台虚拟化（Platform Virtualization）、资源虚拟化（Resource Virtualization）、应用程序虚拟化（Application Virtualization）等。

① 平台虚拟化。它是针对计算机和操作系统的虚拟化，又分为服务器虚拟化和桌面虚拟化。

- 服务器虚拟化是一种通过区分资源的优先次序，将服务器资源分配给最需要它们的工作负载的虚拟化模式，它通过减少为单个工作负载峰值而储备的资源来简化管理和提高效率，如微软（Microsoft）公司的 Hyper-V、思杰（Citrix）公司的 XenServer、威睿（VMware）公司的 ESXi。

- 桌面虚拟化是为提高人对计算机的操控力、降低计算机使用的复杂性，而为用户提供更加方便、适用的使用环境的一种虚拟化模式，如微软公司的 Remote Desktop Services、Citrix 公司的 XenDesktop、VMware 公司的 View。

平台虚拟化主要通过 CPU 虚拟化、内存虚拟化和 I/O 接口虚拟化来实现。

② 资源虚拟化。它是针对特定的计算资源进行的虚拟化，如存储虚拟化、网络资源虚拟化等。存储虚拟化是指把操作系统有机地分布于若干内外存储器上，两者结合成虚拟存储器上。网络资源虚拟化最典型的是网格计算，网格计算通过使用虚拟化技术来管理网络中的数据，并在逻辑上将其作为一个系统呈现给用户，它动态地提供了符合用户和应用程序需求的资源，同时将提供对基础设施的共享和访问的简化功能。当前，一些研究人员提出可利用软件代理技术来实现计算网络空间资源的虚拟化。

③ 应用程序虚拟化。它包括仿真、模拟、解释技术等。Java 虚拟机是典型的在应用层进行虚拟化的虚拟机。基于应用层的虚拟化技术，通过保存用户的个性化计算环境的配置信息，可以在任意计算机上重现用户的个性化计算环境。服务虚拟化是近年研究的一个热点，服务虚拟化可以满足使用户按需快速构建应用的需求，通过服务聚合，可屏蔽服务资源使用的复杂性，使用户更易于直接将业务需求映射到虚拟化的服务资源。现代软件体系结构及其配置的复杂性延长了软件开发生命周期，通过在应用层建立虚拟化的模型，可以提供很好的开发、测试和运行环境。

（4）全虚拟化与半虚拟化

根据虚拟化实现技术的不同，虚拟化可分为全虚拟化（Full Virtualization）和半虚拟化（Para Virtualization）两种，其中，全虚拟化将是未来虚拟化的主流。

① 全虚拟化。全虚拟化也称原始虚拟化技术，用全虚拟化模拟出来的虚拟机中的操作系统是与底层的硬件完全隔离的，虚拟机中所有的硬件资源（包括处理器、内存和外设）都通过虚拟化软件来模拟，支持运行任何理论上可在真实物理平台上运行的操作系统，为虚拟机的配置提供了较大的程序的灵活性。从客户机操作系统看来，完全虚拟化的虚拟平台和现实平台是一样的，客户机操作系统察觉不到其运行在一个虚拟平台上。这样的虚拟平台可以运行现有的操作系统，而无须对操作系统进行任何修改，因此这种技术称为全虚拟化。全虚拟化的运行速度要快于硬件模拟，但是在性能方面不如裸机，因为 Hypervisor 需要占用一些资源。

② 半虚拟化。半虚拟化是另一种类似于全虚拟化的技术，需要修改虚拟机中的操作系统来集成一些虚拟化方面的代码，以减小虚拟化软件的负载。半虚拟化模拟出来的虚拟机整体性能会更好，因为修改后的虚拟机操作系统承载了部分虚拟化软件的工作。其不足之处是，要修改虚拟机的操作系统，用户会感知到使用的环境是虚拟化环境，而且兼容性比较差，因此用户使用时比较麻烦，需要获得集成虚拟化代码的操作系统。

2. 虚拟化技术与云计算的关系

云计算有很广泛的范畴，是中间件、分布式计算（网格计算）、并行计算、效用计算、网络存储、虚拟化和负载均衡等网络技术发展融合的产物。

虚拟化技术不一定必须与云计算相关，如 CPU 虚拟化、内存虚拟化等属于虚拟化技术，但与云计算无关，如图 1.5 所示。

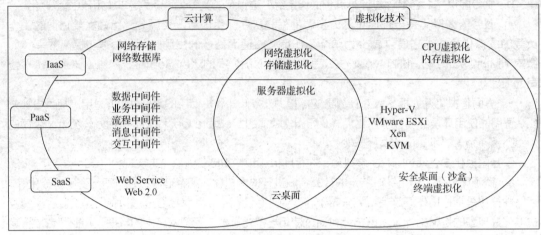

图 1.5 虚拟化技术与云计算的关系

（1）虚拟化技术的特征

① 提高资源利用率。虚拟化技术可实现物理资源和资源池的动态共享，提高资源利用率，特别是针对那些平均需求远低于需要为其提供专用资源的不同负载。

② 降低管理成本。虚拟化技术可通过以下途径提高工作人员的效率：减少必须进行管理的物理资源的数量；隐藏物理资源的部分复杂性；通过实现自动化、获得更好的信息和实现中央管理来简化公共管理任务；实现负载管理自动化。另外，虚拟化技术还可以支持在多个平台上使用公共的工具。

V1-6 虚拟化技术的特征

③ 提高使用灵活性。通过虚拟化技术可实现动态的资源部署和重配置，满足不断变化的业务需求。

④ 提高安全性。虚拟化技术可实现较简单的共享机制所无法实现的隔离和划分，这些特性可实现对数据和服务的可控和安全访问。

⑤ 提高可用性。虚拟化技术可在不影响用户的情况下对物理资源进行删除、升级或改变。

⑥ 提高可扩展性。根据不同的产品，资源分区和汇聚可实现比个体物理资源小得多或大得多的虚拟资源，这意味着用户可以在不改变物理资源配置的情况下进行规模调整。

⑦ 提供互操作性和兼容性。虚拟资源可提供底层物理资源无法提供的与各种接口和协议的互操作性及兼容性。

⑧ 改进资源供应。与个体物理资源单位相比，虚拟化技术能够以更小的单位进行资源分配。

（2）云计算的特征

① 按需自动服务。消费者不需要或很少需要云服务提供商的协助，可以单方面按需获取云端的计算资源。例如，服务器、网络存储等资源是按需自动部署

V1-7 云计算的特征

的，不需要与服务提供商进行人工交互。

② 广泛的网络访问。消费者可以随时随地使用云终端设备接入网络并使用云端的计算资源。常见的云终端设备包括手机、平板电脑、笔记本电脑、掌上电脑和台式计算机等。

③ 资源池化。云端计算资源需要被池化，以便通过多租户形式共享给多个消费者，也只有池化才能根据消费者的需求动态分配或再分配各种物理的和虚拟的资源。消费者通常不知道自己正在使用的计算资源的确切位置，但是在自助申请时允许指定大概的区域范围，例如，在哪个省或者哪个数据中心。

④ 快速、弹性。消费者能方便、快捷地按需获取和释放计算资源。也就是说，需要时能快速获取资源，从而扩展计算能力；不需要时能迅速释放资源，以便降低计算能力，从而减少资源的使用费用。对于消费者来说，云端的计算资源是无限的，可以随时申请并获取任何数量的计算资源。但是一定要消除一个误解，即一个实际的云计算系统不一定是投资巨大的工程，不一定要购买成千上万台计算机，也不一定具备超大规模的运算能力。其实用一台计算机就可以组建一个很小的云端，云端建设方案务必采用可伸缩性策略，刚开始时先采用几台计算机，再根据用户数量规模来增减计算资源。

⑤ 按需按量可计费。消费者使用云端计算资源是要付费的，付费的计量方法有很多，例如，可以根据某类资源（如存储、CPU、内存、网络带宽等）的使用量和时间长短计费，也可以按照使用次数来计费。但不管如何计费，对消费者来说，价格要清楚，计量要明确，而云服务提供商需要监视和控制资源的使用情况，并及时输出各种资源的使用报表，做到供需双方费用结算清楚明白。

3. OpenStack 支持的虚拟化技术

在 OpenStack 环境中，计算服务通过应用程序接口（Application Program Interface，API）服务器来控制虚拟机管理程序，它具备一个抽象层，可以在部署时选择一种虚拟化技术创建虚拟机，向用户提供云服务，OpenStack 支持的虚拟化技术如下。

（1）KVM

基于内核的虚拟机（Kernel-based Virtual Machine，KVM）是通用的开放虚拟化技术，也是 OpenStack 用户使用得较多的虚拟化技术，它支持 OpenStack 的所有特性。

（2）Xen

Xen 是部署快速、安全、开源的虚拟化软件技术，可使多个同样的操作系统或不同操作系统的虚拟机运行在同一主机上，Xen 主要包括服务器虚拟化平台（Xen Server）、云基础架构、管理 XenServer 和 Xen 云计算管理平台（Xen Cloud Platform，XCP）的 API 程序（XenAPI）、基于 Libvirt 的 Xen。OpenStack 通过 XenAPI 支持 XenServer 和 XCP 两种虚拟化技术，但在红帽公司的 Linux（Red Hat Enterprise Linux，RHEL）等平台上，OpenStack 使用的是基于 Libvirt 的 Xen。

（3）容器

容器是在单一 Linux 主机上提供多个隔离的 Linux 环境的操作系统级虚拟化技术，不像基于虚拟管理程序的传统虚拟化技术，容器并不需要运行专用的客户操作系统。目前的容器技术有以下两种。

① Linux 容器（Linux Container，LXC）：提供在单一可控主机节点上支持多个相互隔离的服务器容器同时执行的机制。

② Docker：一个开源的应用容器引擎，让开发者可以把应用以及依赖包打包到一个可移植的容器中，然后发布到任何流行的 Linux 平台上，利用 Docker 也可以实现虚拟化，容器完全使用沙

箱机制，相互之间一般不会有任何接口。

使用 Docker 的目的是尽可能减少容器中运行的程序，甚至减少到只运行单个程序，并且通过 Docker 来管理这个程序。LXC 可以快速兼容所有应用程序和工具以及任意管理和编制层次，以替代虚拟机。

虚拟化管理程序提供了更好的进程隔离，呈现了一个完全的系统。LXC、Docker 除了一些基本隔离之外，并未提供足够的虚拟化管理功能，缺乏必要的安全机制，基于容器的方案无法运行与主机内核不同的其他内核，也无法运行一个完全不同的操作系统。目前，OpenStack 社区对容器的驱动支持还不如虚拟化管理程序。在 OpenStack 项目中，LXC 属于计算服务项目 Nova，通过调用 Libvirt 来实现，Docker 驱动是一种新加入虚拟化管理程序的驱动，目前无法替代虚拟化管理程序。

（4）Hyper-V

Hyper-V 是微软公司推出的企业级虚拟化解决方案，Hyper-V 的设计借鉴了 Xen，管理程序采用微内核的架构，兼顾了安全性和性能要求。Hyper-V 作为一种免费的虚拟化方案，在 OpenStack 中得到了很多支持。

（5）VMware ESXi

VMware 公司提供了业界领先且可靠的服务器虚拟化平台和软件定义计算产品，其中 ESXi 虚拟化平台用于创建和运行虚拟机及虚拟设备，它在 OpenStack 中也得到了支持，但是如果没有 vCenter 和企业级许可，一些 API 的使用会受到限制。

（6）Baremetal 与 Ironic

有些云计算管理平台除了提供虚拟化和虚拟机服务之外，还提供传统的主机服务，在 OpenStack 中，可以对裸金属（Baremetal）与其他部署有虚拟化管理程序的节点通过不同的计算池可用区域（Availability Zone）来管理。Baremetal 是计算服务的后端驱动，与 Libvirt 驱动、VMware 驱动类似，只不过它用来管理没有虚拟化的硬件，主要通过预启动执行环境（Pre-boot eXecution Environment，PXE）和智能平台管理接口（Intelligent Platform Management Interface，IPMI）进行控制管理。

现在 Baremetal 已经由 Ironic 替代，Nova 管理的是虚拟机的生命周期，而 Ironic 管理的是主机的生命周期。Ironic 提供了一系列管理主机的 API，可以对"裸"操作系统的主机进行管理，从主机上架安装操作系统到主机下架维修，可以像管理虚拟机一样管理主机。创建一个 Nova 计算物理节点，只需告诉 Ironic，并自动从镜像模板中加载操作系统到 nova-computer 安装完成即可。Ironic 解决了主机的添加、删除、电源管理、操作系统部署等问题，目标是成为主机管理的成熟解决方案，让 OpenStack 不但可以在软件层面解决云计算问题，而且可以使供应商对应自己的服务器开发 Ironic 插件。

4. 基于 Linux 内核的虚拟化解决方案

KVM 是一种基于 Linux 内核的 x86 硬件平台开源全虚拟化解决方案，也是主流 Linux 虚拟化解决方案，支持广泛的客户机操作系统。KVM 需要 CPU 的虚拟化指令集的支持，如英特尔（Intel）公司的 Intel TV（vmx 指令集）和超威半导体（AMD）公司的 AMD-V（svm 指令集）。

（1）KVM 模块

KVM 模块是一个可加载的内核模块 kvm.ko。KVM 依赖于 x86 硬件架构，因此 KVM 还需要一个处理规范模块。如果使用 Intel 架构，则加载 kvm-intel.ko 模块；如果使用 AMD 模块，则加载 kvm-adm.ko 模块。

KVM 模块负责对虚拟机的虚拟 CPU（vCPU）和内存进行管理及调试，主要任务是初始化

CPU，进入虚拟化模式，将虚拟机运行在虚拟化模式下，并对虚拟机的运行提供一定的支持。

至于虚拟机的外部设备交互，如果是真实的物理硬件设备，则利用 Linux 操作系统内核来管理；如果是虚拟的外部设备，则借助快速仿真（Quick Emulator，QEMU）来处理。

由此可见，KVM 本身只关注虚拟机的调试和内存管理，是一个轻量级的 Hypervisor。很多 Linux 发行版集成 KVM 作为虚拟化解决方案使用，CentOS 也不例外。

（2）QEMU

KVM 模块本身无法作为 Hypervisor 模拟出一个完整的虚拟机，用户也不能直接对 Linux 操作系统内核进行操作，因此需要借助其他软件来进行，QEMU 就是 KVM 所需要的这样一款软件。

QEMU 并非 KVM 的一部分，而是一款开源的虚拟机软件，与 KVM 不同，作为一个宿主型的 Hypervisor，就算没有 KVM，QEMU 也可以通过模拟来创建和管理虚拟机，但因为是纯软件实现的，所以其性能较低。QEMU 的优点是，在支持 QEMU 编译运行的平台上就可以实现虚拟机的功能，甚至虚拟机可以与主机不在同一个架构。KVM 在 QEMU 的基础上进行了修改，虚拟机运行期间，QEMU 会通过 KVM 模块提供的系统调用进入内核，由 KVM 模块负责将虚拟机置于处理器的特殊模式运行，遇到虚拟机进行 I/O 操作时，KVM 模块转交给 QEMU 解析和模拟这些设备。

QEMU 使用 KVM 模块的虚拟化功能，为自己的虚拟机提供硬件虚拟化的加速，从而极大地提高了虚拟机的性能。除此之外，虚拟机的配置和创建、虚拟机运行依赖的虚拟设备、虚拟机运行时的用户操作环境和交互以及一些针对虚拟机的特殊技术（如动态迁移），都是由 QEMU 自己实现的。

KVM 虚拟机的创建和运行是一个用户空间的 QEMU 程序和内核空间的 KVM 模块相互配合的过程。KVM 模块作为整个虚拟化环境的核心而工作在系统空间，负责 CPU 和内存的调试等；QEMU 作为模拟器而工作在用户空间，负责虚拟机 I/O 模拟等。

（3）KVM 架构

从上面的分析来看，KVM 作为 Hypervisor 主要包括两个重要的组成部分：一个是 Linux 操作系统的 KVM 模块，主要负责虚拟机的创建、虚拟内存的分配、vCPU 寄存器的读写以及 vCPU 的运行；另一个是提供硬件仿真的 QEMU，主要用于模拟虚拟机的用户空间组件、提供 I/O 设备模型和访问外设的途径。KVM 的基本架构如图 1.6 所示。

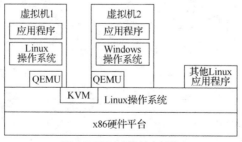

图 1.6　KVM 的基本架构

在 KVM 模型中，每一个虚拟机都是一个由 Linux 调度程序管理的标准进程，可以在用户空间启动客户机操作系统。一个普通的 Linux 进程有两种运行模式——内核和用户，而 KVM 增加了第三种模式，即客户模式，客户模式又有自己的内核和用户模式。当新的虚拟机在 KVM 上启动时，它就成为主机操作系统的一个进程，因此可以像其他进程一样调度，但与传统的 Linux 进程不一样，客户端被 Hypervisor 标识为处于 Guest 模式（独立于内核和用户模式），每个虚拟机都是通过 /dev/kvm 设备映射的，它们拥有自己的虚拟地址空间，该空间映射到主机内核的物理地址空间。如前所述，KVM 使用层硬件的虚拟化支持来提供完整的（原生）虚拟化，I/O 请求通过主机内核映射到主机上（Hypervisor）执行的 QEMU 进程上。

（4）KVM 虚拟磁盘（镜像）文件格式

在 KVM 中往往使用镜像（Image）这类术语来表示虚拟磁盘，其主要有以下 3 种文件格式。

① RAW。原始的文件格式，它直接将文件系统的存储单元分配给虚拟机使用，采取直读直写的策略。该文件格式实现简单，不支持诸如压缩、快照、加密和复制等特性。

② QCOW2。QEMU 引入的镜像文件格式，也是目前 KVM 默认的格式。QCOW2 格式的文件存储数据的基本单元是簇（Cluster），每一簇由若干个数据扇区组成，每个数据扇区的大小是 512 字节。在 QCOW2 中，要定位镜像文件的簇，需要经过两次地址查询操作，QCOW2 根据实际需要来决定占用空间的大小，而且支持更多的主机文件格式。

③ QED。QCOW2 的一种改进格式，QED 的存储、定位、查询方式以及数据块大小与 QCOW2 的一样，它的目的是克服 QCOW2 文件格式的一些缺点，提高性能，不过目前还不够成熟。

如果需要使用虚拟机快照，则需要选择 QCOW2 文件格式，对于大规模数据的存储，可以选择 RAW 文件格式。使用 QCOW2 文件格式只能增加容量，不能减少容量，而使用 RAW 文件格式可以增加或者减少容量。

5. Libvirt 套件

仅有 KVM 模块和 QEMU 组件是不够的，为了使 KVM 的整个虚拟环境易于管理，还需要 Libvirt 服务和基于 Libvirt 开发出来的管理工具。

Libvirt 是一个软件集合，是一套为方便管理平台虚拟化技术而设计的开源代码的 API、守护进程和管理工具。它不仅提供了对虚拟机的管理，还提供了对虚拟网络和存储的管理。Libvirt 最初是为 Xen 虚拟化平台设计的一套 API，目前支持其他多种虚拟化平台，如 KVM、ESX 和 QEMU 等。在 KVM 解决方案中，QEMU 用来进行平台模拟，面向上层管理和操作；Libvirt 用来管理 KVM，面向下层管理和操作。Libvirt 架构如图 1.7 所示。

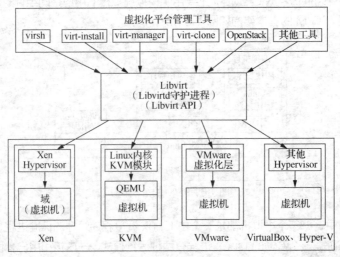

图 1.7　Libvirt 架构

Libvirt 是目前使用广泛的虚拟机管理应用程序接口，一些常用的虚拟机管理工具（如 virsh）和云计算框架平台（如 OpenStack）都是在底层使用 Libvirt 的 API。

Libvirt 包括两部分，一部分是服务（守护进程名为 Libvirtd），另一部分是 API。作为一个运行在主机上的服务端守护进程，Libvirtd 为虚拟化平台及其虚拟机提供本地和远程的管理功能，基于 Libvirt 开发出来的管理工具可通过 Libvirtd 服务来管理整个虚拟化环境。也就是说，Libvirtd 在管理工具和虚拟化平台之间起到了桥梁的作用。Libvirt API 是一系列标准的库文件，给多种虚拟化平台提供了统一的编程接口，相当于管理工具一定要基于 Libvirt 的标准接口来进行开发，而开发完成后的工具可支持多种虚拟化平台。

1.2.3 OpenStack 概述

OpenStack 是一个旨在为公有云及私有云的建设与管理提供软件的开源项目。它的社区拥有众多企业及 1350 位开发者，这些企业与个人将 OpenStack 作为基础设施即服务资源的通用前端。OpenStack 项目的首要任务是简化云的部署过程并为其带来良好的可扩展性。

OpenStack 为私有云和公有云提供可扩展的弹性的云计算服务，项目目标是提供实施简单、丰富、标准统一、可大规模扩展的云计算管理平台。

1. OpenStack 的起源

OpenStack 是一个开源的云计算管理平台项目，是一系列软件开源项目的组合，是美国国家航空航天局（National Aeronautics and Space Administration，NASA）和 Rackspace（美国的一家云计算厂商）在 2010 年 7 月共同发起的一个项目，旨在为公有云和私有云提供软件的开源项目，由 Rackspace 贡献存储源码（Swift）、NASA 贡献计算源码（Nova）。

V1-8 OpenStack 的起源

经过几年的发展，OpenStack 现已成为一个广泛使用的业内领先的开源项目，提供部署私有云及公有云的操作平台和工具集，并且在许多大型企业支撑核心生产业务。

OpenStack 示意如图 1.8 所示。OpenStack 项目旨在提供开源的云计算解决方案以简化云的部署过程，其实现类似于 AWS EC2 和 S3 的 IaaS。其主要应用场合包括 Web 应用、大数据、电子商务、视频处理与内容分发、大吞吐量计算、容器优化、主机托管、公有云和数据库等。

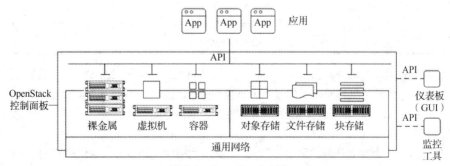

图 1.8 OpenStack 示意

Open 意为开放，Stack 意为堆栈或堆叠，OpenStack 是一系列软件开源项目的组合，包括若干项目，每个项目都有自己的代号（名称），包括不同的组件，每个组件又包括若干服务，一个服务意味着运行的一个进程。这些组件部署灵活，支持水平扩展，具有伸缩性，支持不同规模的云计算管理平台。

OpenStack 最初仅包括 Nova 和 Swift 两个项目，现在已经有数十个项目，其主要项目如表 1.1 所示。这些项目相互关联，协同管理各类计算、存储和网络资源，提供云计算服务。

表 1.1 OpenStack 的主要项目

项目名称	服务	功能
Horizon	仪表板（Dashboard）	提供一个与 OpenStack 服务交互的基于 Web 的自服务网站，让最终用户和运维人员都可以完成大多数的操作，例如，启动虚拟机、分配 IP 地址、动态迁移等
Nova	计算（Compute）	部署与管理虚拟机并为用户提供虚拟机服务，管理 OpenStack 环境中计算实例的生命周期，按需响应，包括生成、调试、回收虚拟机等操作
Neutron	网络（Networking）	为 OpenStack 其他服务提供网络连接服务，为用户提供 API 定义网络和接入网络，允许用户创建自己的虚拟网络并连接各种网络设备接口。它提供基于插件的架构，支持众多的网络提供商和技术
Swift	对象存储（Object Storage）	一套用于在大规模可扩展系统中通过内置冗余及高容错机制实现对象存储的系统，允许进行存储或者检索文件。可为 Glance 提供镜像存储，为 Cinder 提供卷备份服务
Cinder	块存储（Block Storage）	为运行实例提供稳定的数据块存储服务，它的插件驱动架构有利于块设备的创建和管理，如创建卷、删除卷，在实例上挂载和卸载卷等
Keystone	身份服务（Identity Service）	为 OpenStack 其他服务提供身份验证、服务规则和服务令牌的功能，管理 Domains、Projects、Users、Groups、Roles
Glance	镜像服务（Image Service）	一套虚拟机镜像查找及检索系统，支持多种虚拟机镜像格式（AKI、AMI、ARI、ISO、QCOW2、RAW、VDI、VHD、VMDK），有创建镜像、上传镜像、删除镜像、编辑镜像基本信息的功能
Ceilometer	测量（Metering）	像一个漏斗一样，能把 OpenStack 内部发生的几乎所有的事件都收集起来，并为计费和监控以及其他服务提供数据支撑
Heat	部署编排（Orchestration）	提供了一种通过模板定义的协同部署方式，实现云基础设施软件运行环境（计算、存储和网络资源）的自动化部署
Trove	数据库（Database）	为用户在 OpenStack 的环境中提供可扩展和可靠的关系及非关系数据库引擎服务
Sahara	数据处理（Data Processing）	为用户提供简单部署 Hadoop 集群的能力，如通过简单配置迅速将 Hadoop 集群部署起来

作为免费的开源软件项目，OpenStack 由一个名为 OpenStack Community 的社区开发和维护，来自世界各地的云计算开发人员和技术人员共同开发、维护 OpenStack 项目。与其他开源的云计算软件相比，OpenStack 在控制性、兼容性、灵活性方面具备优势，它有可能成为云计算领域的行业标准。

（1）控制性。作为完全开源的平台，OpenStack 为模块化的设计，提供相应的 API，方便与第三方技术集成，从而满足自身业务需求。

（2）兼容性。OpenStack 兼容其他公有云，方便用户进行数据迁移。

（3）可扩展性。OpenStack 采用模块化的设计，支持各主流发行版本的 Linux，可以通过横向扩展增加节点、添加资源。

（4）灵活性。用户可以根据自己的需要建立基础设施，也可以轻松地为自己的群集扩大规模。OpenStack 项目采用 Apache2 许可，意味着第三方厂商可以重新发布源代码。

（5）行业标准。众多 IT 领军企业都加入了 OpenStack 项目，意味着 OpenStack 在未来可能成为云计算行业标准。

2. OpenStack 版本演变

2010 年 10 月，OpenStack 第 1 个正式版本发布，其代号为 Austin。第 1 个正式版本仅有 Swift 和 Nova（对象存储和计算）两个项目。起初计划每隔几个月发布一个全新的版本，并且以 26 个英文字母为首字母，以 A 到 Z 的顺序命名后续版本。2011 年 9 月，其第 4 个版本 Diablo 发布时，定为约每半年发布一个版本，分别是当年的春秋两季，每个版本不断改进，吸收新技术，实现新概念。

2020 年 5 月 13 日，其发布了第 21 个版本，即 Ussuri，如今该版本已经更加稳定、更加强健。近几年，Docker、Kubernetes、Serverless 等新技术兴起，而 OpenStack 的关注点不再是"谁是龙头"，而是"谁才是最受欢迎的技术"。

OpenStack 不受任何一家厂商的绑定，灵活自由。当前可以认为它是云解决方案的首选方案。当前多数私有云用户转向 OpenStack，因为它使用户摆脱了对单个公有云的过多依赖。实际上，OpenStack 用户经常依赖于公有云，例如，对于 Amazon Web Services（AWS）、Microsoft Azure 或 Google Compute Engine，多数用户基础架构是由 OpenStack 驱动的。

尽管 OpenStack 从诞生到现在已经日渐成熟，基本上能够满足云计算用户的大部分需求，但随着云计算技术的发展，OpenStack 必然需要不断地完善。OpenStack 已经逐渐成为市场上一个主流的云计算管理平台解决方案。

3. OpenStack 的架构

在学习 OpenStack 的部署和运维之前，应当熟悉其架构和运行机制。OpenStack 作为一个开源、可扩展、富有弹性的云操作系统，其架构设计主要参考了 AWS 云计算产品，通过模块的划分和模块的功能协作，设计的基本原则如下。

① 按照不同的功能和通用性划分不同的项目，拆分子系统。
② 按照逻辑计划、规范子系统之间的通信。
③ 通过分层设计整个系统架构。
④ 不同功能子系统间提供统一的 API。

（1）OpenStack 的逻辑架构

OpenStack 的逻辑架构如图 1.9 所示。此图展示了 OpenStack 云计算管理平台各模块（仅给出主要服务）协同工作的机制和流程。

OpenStack 通过一组相关的服务提供一个基础设施即服务的解决方案，这些服务以虚拟机为中心。虚拟机主要是由 Nova、Glance、Cinder 和 Neutron 这 4 个核心模块进行交互的结果。Nova 为虚拟机提供计算资源，包括 vCPU、内存等；Glance 为虚拟机提供镜像服务，安装操作系统的运行环境；Cinder 提供存储资源，类似传统计算机的磁盘或卷；Neutron 为虚拟机提供网络配置以及访问云计算管理平台的网络通道。

云计算管理平台用户在经过 Keystone 认证授权后，通过 Horizon 创建虚拟机服务。创建过程包括利用 Nova 服务创建虚拟机实例，虚拟机实例采用 Glance 提供的镜像服务，使用 Neutron 为新建的虚拟机分配 IP 地址，并将其纳入虚拟网络，之后通过 Cinder 创建的卷为虚拟机挂载存储块。整个过程都在 Ceilometer 的监控下，Cinder 产生的卷（Volume）和 Glance 提供的镜像可以通过 Swift 的对象存储机制进行保存。

Horizon、Ceilometer、Keystone 分别提供访问、监控、身份认证（权限）等功能，Swift 提供对象存储功能，Heat 实现应用系统的自动化部署，Trove 用于部署和管理各种数据库，Sahara 提供大数据处理框架，而 Ironic 提供裸金属云服务。

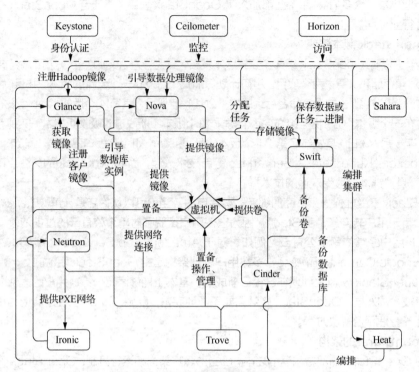

图 1.9　OpenStack 的逻辑架构

云计算管理平台用户通过 nova-api 等来与其他 OpenStack 服务交互，而这些 OpenStack 服务守护进程通过消息总线（动作）和数据库（信息）来执行 API 请示。

消息队列为所有守护进程提供一个中心的消息机制，消息的发送者和接收者相互交换任务或数据进行通信，协同完成各种云计算管理平台功能。消息队列对各个服务进程解耦，所有进程可以任意地进行分布式部署，协同工作。

（2）OpenStack 的物理架构

OpenStack 是分布式系统，必须从逻辑架构映射到具体的物理架构，将各个项目和组件以一定的方式安装到实际服务器节点上，部署到实际的存储设备上，并通过网络将它们连接起来。这就形成了 OpenStack 的物理架构。

OpenStack 的部署分为单节点部署和多节点部署两种类型。单节点部署就是将所有的服务和组件都部署在一个物理节点上，通常用于学习、验证、测试或者开发；多节点部署就是将服务和组件分别部署在不同的物理节点上。OpenStack 的多节点部署如图 1.10 所示，常见的节点类型有控制节点（Control Node）、网络节点（Network Node）、计算节点（Compute Node）和存储节点（Storage Node）。

① 控制节点。控制节点又称管理节点，可安装并运行各种 OpenStack 控制服务，负责管理、节制其余节点，执行虚拟机建立、迁移、网络分配、存储分配等任务。OpenStack 的大部分服务运行在控制节点上。

- 支持服务。数据库服务，如 MySQL 数据库；消息队列服务，如 RabbitMQ。
- 基础服务。运行 Keystone 认证服务、Glance 镜像服务、Nova 计算服务的管理组件，以及 Neutron 网络服务的管理组件和 Horizon 仪表板。

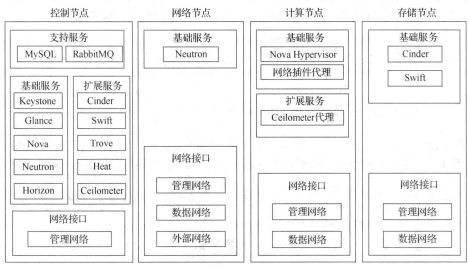

图 1.10 OpenStack 的多节点部署

- 扩展服务。运行 Cinder 块存储服务、Swift 对象存储服务、Trove 数据库服务、Heat 编排服务和 Ceilometer 计量服务的部分组件，这对于控制节点来说是可选的。

控制节点一般只需要一个网络接口，用于通信和管理各个节点。

② 网络节点。网络节点可实现网关和路由的功能，它主要负责外部网络与内部网络之间的通信，并将虚拟机连接到外部网络。网络节点仅包含 Neutron 基础服务，Neutron 负责管理私有网段与公有网段的通信、虚拟机网络之间的通信拓扑，以及虚拟机上的防火墙等。

网络节点通常需要 3 个网络接口，分别用来与控制节点进行通信、与除控制节点外的计算节点和存储节点进行通信、与外部的虚拟机和相应网络进行通信。

③ 计算节点。计算节点是实际运行虚拟机的节点，主要负责虚拟机的运行，为用户创建并运行虚拟机，为虚拟机分配网络。计算节点通常包括以下服务。

- 基础服务。Nova 计算服务的虚拟机管理器组件（Hypervisor），提供虚拟机的创建、运行、迁移、快照等各种围绕虚拟机的服务，并提供 API 与控制节点的对接，由控制节点下发任务，默认计算服务使用的 Hypervisor 是 KVM；网络插件代理，用于将虚拟机实例连接到虚拟网络，通过安全组件为虚拟机提供防火墙服务。
- 扩展服务。Ceilometer 计量服务代理，提供计算节点的监控代理，将虚拟机的情况反馈给控制点。

虚拟机可以部署多个计算节点，一个计算节点至少需要两个网络接口：一个与控制节点进行通信，受控制节点统一配置；另一个与网络节点及存储节点进行通信。

④ 存储节点。存储节点负责对虚拟机的外部存储进行管理等，即为计算节点的虚拟机提供持久化卷服务，这种节点存储需要的数据，包括磁盘镜像、虚拟机持久性卷。存储节点包含 Cinder 和 Swift 等基础服务，可根据需要安装共享文件服务。

块存储和对象存储可以部署在同一个存储节点上，也可以分别部署，无论采用哪种方式，都可以部署多个存储节点。

最简单的网络连接存储节点只需要一个网络接口，可直接使用管理网络在计算节点和存储节点之间进行通信。在生产环境中，存储节点最少需要两个网络接口：一个连接管理网络，与控制节点进行通信，接受控制节点下发的任务，受控制节点统一调配；另一个连接专门的存储网络（数据网络），与计算节点和网络节点进行通信，完成控制节点下发的各类数据传输任务。

1.3 项目实施

1.3.1 VMware Workstation 安装

虚拟机软件有很多，本书选用 VMware Workstation 软件。VMware Workstation 是一款功能强大的桌面虚拟机软件，可以在单一桌面上同时运行不同操作，并完成开发、调试、部署等。

（1）下载 VMware-workstation-full-16.1.2-17966106 软件安装包，双击安装文件，弹出 VMware 安装主界面，如图 1.11 所示。

（2）单击"下一步"按钮，弹出 VMware 安装界面，如图 1.12 所示。

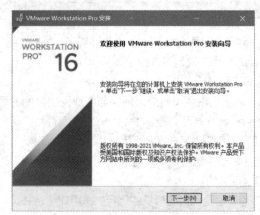

图 1.11 VMware 安装主界面

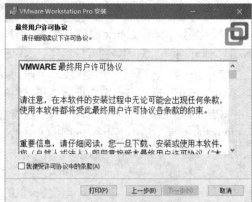

图 1.12 VMware 安装界面

（3）选中"我接受许可协议中的条款"复选框，如图 1.13 所示。

（4）单击"下一步"按钮，弹出 VMware 自定义安装界面，如图 1.14 所示。

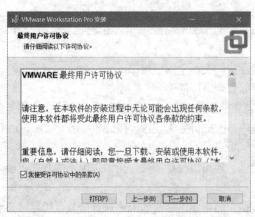

图 1.13 接受 VMware 许可协议中的条款

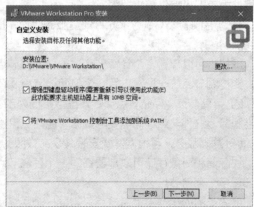

图 1.14 VMware 自定义安装界面

（5）选中图 1.14 中的复选框，单击"下一步"按钮，弹出 VMware 用户体验设置界面，如图 1.15 所示。

（6）保留默认设置，单击"下一步"按钮，弹出 VMware 快捷方式界面，如图 1.16 所示。

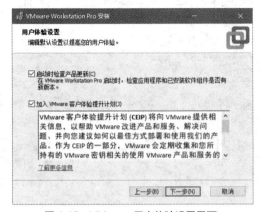

图 1.15　VMware 用户体验设置界面

图 1.16　VMware 快捷方式界面

（7）保留默认设置，单击"下一步"按钮，弹出 VMware 准备安装界面，如图 1.17 所示。

（8）单击"安装"按钮，开始安装，弹出 VMware 正在安装界面，如图 1.18 所示。

图 1.17　VMware 准备安装界面

图 1.18　VMware 正在安装界面

（9）单击"完成"按钮，完成安装，弹出 VMware 安装向导已完成界面，如图 1.19 所示。

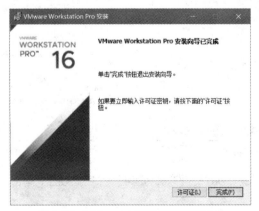

图 1.19　VMware 安装向导已完成界面

1.3.2 虚拟机安装

（1）从 CentOS 官网下载 Linux 发行版的 CentOS 安装包，本书使用的下载文件为"CentOS-7-x86_64-DVD-1810.iso"，当前版本为 7.6.1810。

（2）双击桌面上的 VMware Workstation Pro 图标，如图 1.20 所示，打开软件。

（3）启动后会弹出 VMware Workstation 界面，如图 1.21 所示。

图 1.20　VMware Workstation Pro 图标

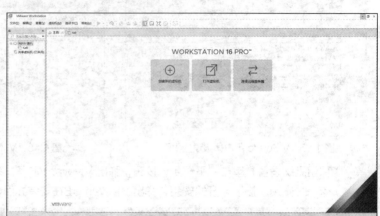

图 1.21　VMware Workstation 界面

（4）使用新建虚拟机向导，安装虚拟机，默认选中"典型(推荐)"单选按钮，单击"下一步"按钮，如图 1.22 所示。

（5）安装客户机操作系统，可以选中"安装程序光盘"或选中"安装程序光盘映像文件(iso)"单选按钮，并浏览选中相应的 ISO 文件，也可以选中"稍后安装操作系统"单选按钮。本次选中"稍后安装操作系统"单选按钮，并单击"下一步"按钮，如图 1.23 所示。

图 1.22　新建虚拟机向导

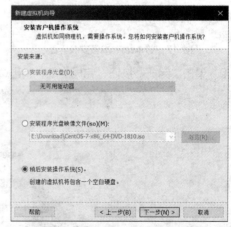

图 1.23　安装客户机操作系统

（6）选择客户机操作系统，创建的虚拟机将包含一个空白硬盘，单击"下一步"按钮，如图 1.24 所示。

（7）命名虚拟机，选择系统文件安装位置，单击"下一步"按钮，如图 1.25 所示。

图 1.24 选择客户机操作系统

图 1.25 命名虚拟机

（8）指定磁盘容量，并单击"下一步"按钮，如图 1.26 所示。
（9）已准备好创建虚拟机，如图 1.27 所示。

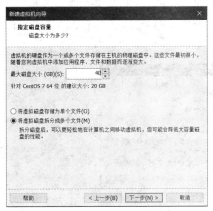
图 1.26 指定磁盘容量

图 1.27 已准备好创建虚拟机

（10）单击"自定义硬件"按钮，进行虚拟机硬件相关信息配置，如图 1.28 所示。

图 1.28 虚拟机硬件相关信息配置

（11）单击"关闭"按钮，虚拟机初步配置完成，如图 1.29 所示。

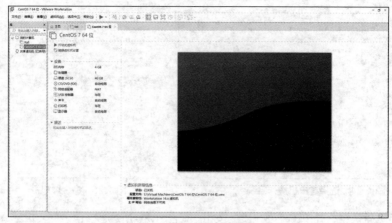

图 1.29　虚拟机初步配置完成

（12）进行虚拟机设置，选择"CD/DVD(IDE)"选项，选中"使用 ISO 映像文件"单选按钮，单击"浏览"按钮，选择 ISO 镜像文件"CentOS-7-x86_64-DVD-1810.iso"，单击"确定"按钮，如图 1.30 所示。

图 1.30　选择 ISO 镜像文件

（13）安装 CentOS，如图 1.31 所示。

（14）设置语言，选择"中文"→"简体中文（中国）"选项，如图 1.32 所示，单击"继续"按钮。

（15）进行安装信息摘要的配置，如图 1.33 所示，可以进行"安装位置"配置，自定义分区，也可以进行"网络和主机名配置"，单击"保存"按钮，返回安装信息摘要的配置界面。

（16）进行软件选择的配置，可以安装桌面化 CentOS。可以选择安装 GNOME 桌面，并选择相关环境的附加选项，如图 1.34 所示。

项目 1
OpenStack 云计算基础

图 1.31　CentOS 安装

图 1.32　设置语言

图 1.33　安装信息摘要的配置

图 1.34　软件选择的配置

（17）单击"完成"按钮，返回 CentOS 7 安装界面，继续进行安装，配置用户设置，如图 1.35 所示。

（18）安装 CentOS 7 的时间稍长，请耐心等待。可以选择"ROOT 密码"选项，进行 ROOT 密码设置，设置完成后单击"完成"按钮，返回安装界面，如图 1.36 所示。

图 1.35　配置用户设置

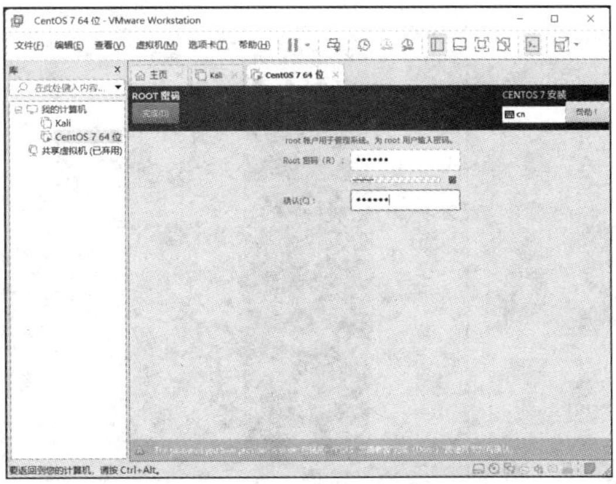

图 1.36　ROOT 密码设置

（19）CentOS 7 安装完成，如图 1.37 所示。

（20）单击"重启"按钮，系统重启后，进入系统，可以进行系统初始设置，如图 1.38 所示。

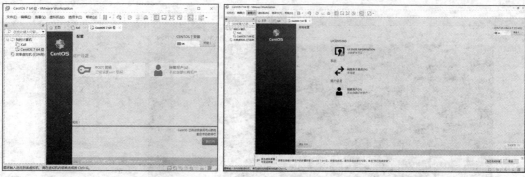

图 1.37　CentOS 7 安装完成　　　　　图 1.38　系统初始设置

（21）单击"退出"按钮，弹出 CentOS 7 Linux EULA 许可协议界面，选中"我同意许可协议"复选框，如图 1.39 所示。

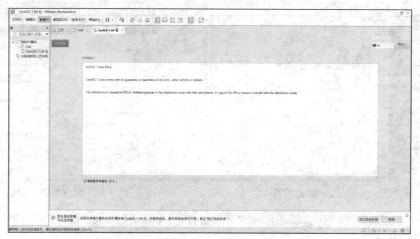

图 1.39　CentOS 7 Linux EULA 许可协议界面

（22）单击"完成"按钮，弹出初始设置界面，如图 1.40 所示。

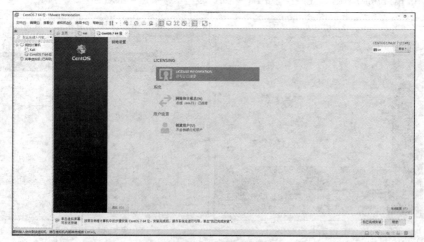

图 1.40　初始设置界面

（23）单击"完成配置"按钮，弹出欢迎界面，选择语言为汉语，如图 1.41 所示。

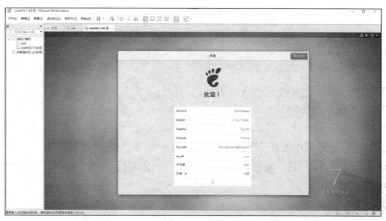

图 1.41 选择语言为汉语

(24)单击"前进"按钮,弹出时区界面,在查找地址栏中输入"上海",选择"上海,上海,中国"选项。

(25)单击"前进"按钮,弹出在线账号界面,如图 1.42 所示。

图 1.42 在线账号界面

(26)单击"跳过"按钮,弹出准备好了界面,如图 1.43 所示。

图 1.43 准备好了界面

课后习题

1. 选择题

（1）云计算服务模式不包括（　　）。
　　A. IaaS　　　　B. PaaS　　　　C. SaaS　　　　D. LaaS

（2）【多选】从服务方式角度可以把云计算分为（　　）3类。
　　A. 公有云　　　B. 私有云　　　C. 金融云　　　D. 混合云

（3）PaaS 是指（　　）。
　　A. 基础设施即服务　　　　　　B. 平台即服务
　　C. 软件即服务　　　　　　　　D. 安全即服务

（4）【多选】云计算的生态系统主要涉及（　　）。
　　A. 硬件　　　　B. 软件　　　　C. 服务　　　　D. 网络

（5）【多选】关于 OpenStack 云计算管理平台，以下说法正确的是（　　）。
　　A. OpenStack 兼容其他公有云，方便用户进行数据迁移
　　B. OpenStack 采用了模块化的设计，支持 Linux 的各主流发行版，可以通过横向扩展增加节点、添加资源
　　C. OpenStack 项目采用了 Apache2 许可，意味着第三方厂商可以重新发布源代码
　　D. 众多 IT 领军企业都加入了 OpenStack 项目，意味着 OpenStack 在未来可能成为云计算行业标准

2. 简答题

（1）简述云计算的定义。
（2）简述云计算的服务模式。
（3）简述云计算的部署类型。
（4）简述云计算的生态系统。
（5）简述虚拟化体系结构。
（6）简述云计算的特征。
（7）简述 OpenStack 的起源以及项目的主要组成部分。
（8）简述 OpenStack 的逻辑架构。

项目2
OpenStack安装与部署

【学习目标】

- 掌握Linux相关知识。
- 理解云计算管理平台部署需求与规划。
- 掌握使用Packstack一键部署OpenStack的步骤与方法。

2.1 项目陈述

OpenStack 的安装部署是一个难题，它的组件众多，安装部署非常烦琐。对于刚刚接触 OpenStack 的初学者而言，安装云计算管理平台更是难上加难，这在很大程度上提高了学习 OpenStack 云计算的技术门槛。因此，可以利用 Red Hat 推出的远程桌面管理（Remote Desktop Organizer，RDO）软件的 Packstack 安装工具快速部署一个 OpenStack 云测试平台，为学习和测试 OpenStack 组件提供平台运行环境。本项目主要帮助读者掌握搭建 OpenStack 云计算管理平台的环境设计及系统部署，包括硬件基本需求、OpenStack 云计算管理平台搭建所需的软件包、一键部署 OpenStack 云计算管理平台的步骤与方法。

2.2 必备知识

2.2.1 Linux 相关知识

Linux 操作系统是一种类 UNIX 操作系统，UNIX 是一种主流经典的操作系统。Linux 源于 UNIX，Linux 操作系统是 UNIX 在计算机上的完整实现。UNIX 操作系统是 1969 年由肯·汤普森（Ken Thompson）在美国贝尔实验室开发的一种操作系统，1973 年，肯·汤普森与丹尼斯·里奇（Dennis Ritchie）一起用 C 语言重写了 UNIX 操作系统，大幅增加了其可移植性。由于其具有良好而稳定的性能，在随后几十年中人们对它做了不断的改进，其迅速发展并在计算机领域得到了广泛的应用。

1. Vi 编辑器与 Vim 编辑器的使用

可视化接口（Visual interface，Vi）也称可视化界面，它为用户提供了一个全屏幕的窗口编辑器，窗口中一次可以显示一屏的编辑内容，并可以上下屏滚动。Vi 编辑器是所有 UNIX 和 Linux 操作系统中标准的编辑器，类似于 Windows 操作系统中的 Notepad++编辑器。因为在任何版本的 UNIX 和 Linux 操作系统中，Vi 编辑器都是完全相同的，所以在其他任何介绍 Vi 编辑器的地方都能进一步了解它。Vi 编辑器也是 Linux 系统中最基本的文本编辑器，学会它后，可以在 Linux 系统的世界中畅通无阻，尤其是在终端中。

Vim（Visual interface improved，Vim）编辑器可以看作 Vi 编辑器的改进升级版。Vi 编辑器和 Vim 编辑器都是 Linux 操作系统中的编辑器，不同的是，Vim 编辑器比较高级。Vi 编辑器用于文本编辑，但 Vim 编辑器更适用于面向开发者的云端开发平台。

Vim 编辑器可以执行输出、移动、删除、查找、替换、复制、粘贴、撤销、块操作等众多文件操作，用户可以根据自己的需要对其进行定制，这是其他编辑器没有的功能。Vim 编辑器不是一个排版器，它不像 Microsoft Word 或 WPS 那样可以对字体、格式、段落等进行编排，它只是一个文件编辑器。Vim 编辑器是全屏幕文件编辑器，没有菜单，只有命令。

在命令行中执行命令#vim filename。如果 filename 已经存在，则 filename 被打开且显示其内容；如果 filename 不存在，则 Vim 编辑器在第一次存盘时自动在硬盘中新建 filename 文件。

Vim 编辑器有 3 种基本工作模式：命令模式、编辑模式、末行模式。考虑到具体需要，用户可以采用状态切换的方法实现工作模式的转换。切换只是习惯性的问题，一旦熟练地使用 Vim 编辑器，就会觉得其使用起来非常方便。

V2-1　Vi、Vim 编辑器的使用

（1）命令模式

命令模式（其他模式→Esc）是用户进入 Vim 编辑器的初始模式，在此模式中，用户可以输入 Vim 命令，让 Vim 编辑器完成不同的工作任务，如光标移动、复制、粘贴、删除等。也可以从其他模式返回到命令模式，在编辑模式下按 Esc 键或在末行模式下输入错误命令都会返回到命令模式。常用的 Vim 命令模式的光标移动命令如表 2.1 所示，Vim 命令模式的复制和粘贴命令如表 2.2 所示，Vim 命令模式的删除命令如表 2.3 所示，Vim 命令模式的撤销与恢复命令如表 2.4 所示。

表 2.1　常用的 Vim 命令模式的光标移动命令

命令	功能说明
gg	将光标移动到屏的首行
G	将光标移动到屏的尾行
w 或 W	将光标移动到下一个单词
H	将光标移动到该屏幕的顶端
M	将光标移动到该屏幕的中间
L	将光标移动到该屏幕的底端
h(←)	将光标向左移动一格
l(→)	将光标向右移动一格
j(↓)	将光标向下移动一格
k(↑)	将光标向上移动一格
0(Home)	数字 0，将光标移动到行首
$(End)	将光标移动到行尾
PageUp/PageDown	（Ctrl+b/Ctrl+f）上下翻屏

表 2.2 Vim 命令模式的复制和粘贴命令

命令	功能说明
yy 或 Y（大写）	复制光标所在的整行
3yy 或 y3y	复制 3 行（含当前行，后 3 行），如复制 5 行，即 5yy 或 y5y
y1G	复制至行文件首
yG	复制至行文件尾
yw	复制一个单词
y2w	复制两个字
p（小写）	粘贴到光标的后（下）面，如果复制的是整行，则粘贴到光标所在行的下一行
P（大写）	粘贴到光标的前（上）面，如果复制的是整行，则粘贴到光标所在行的上一行

表 2.3 Vim 命令模式的删除命令

命令	功能说明
dd	删除当前行
3dd 或 d3d	删除 3 行（含当前行，后 3 行），如删除 5 行，即 5dd 或 d5d
d1G	删除至文件首
dG	删除至文件尾
D 或 d$	删除至行尾
dw	删除至词尾
ndw	删除后面的 n 个词

表 2.4 Vim 命令模式的撤销与恢复命令

命令	功能说明
u（小写）	取消上一个更改（常用）
U（大写）	取消一行内的所有更改
Ctrl+r	重做一个动作（常用），通常与"u"配置使用，将会为编辑提供很多方便
.	这就是小数点，意思是重复前一个动作，如果想要重复删除、复制、粘贴等操作，则按下小数点即可（常用）

（2）编辑模式

在编辑模式（命令模式→a/A、i/I 或 o/O）下，可对编辑的文件添加新的内容并修改内容。这是该模式的唯一功能，即文本输入。要进入该模式，可按 a/A、i/I 或 o/O 键。Vim 编辑模式命令如表 2.5 所示。

表 2.5 Vim 编辑模式命令

命令	功能说明
a（小写）	在光标之后插入内容
A（大写）	在光标当前行的末尾插入内容
i（小写）	在光标之前插入内容
I（大写）	在光标当前行的开始部分插入内容
o（小写）	在光标所在行的下面新增一行
O（大写）	在光标所在行的上面新增一行

（3）末行模式

末行模式（命令模式→":"、"/"或"?"）主要用来实现一些文字编辑辅助功能，如查找、替换、文件保存等，在命令模式中输入":"字符，就可以进入末行模式，若完成了输入的命令或命令出错，则会退出 Vim 编辑器或返回到命令模式。Vim 末行模式命令如表 2.6 所示。按 Esc 键可返回到命令模式。

表 2.6 Vim 末行模式命令

命令	功能说明
ZZ（大写）	保存当前文件并退出
：wq 或：x	保存当前文件并退出
：q	结束 Vim 编辑器程序，如果文件有过修改，则必须先存储文件
：q!	强制结束 Vim 编辑器程序，修改后的文件不会存储
：w[文件路径]	保存当前文件，且保存为另一个文件（类似另存为新文件）
：r[filename]	在编辑的数据中，读入另一个文件的数据，即将 filename 这个文件的内容加到光标所在行的后面
：!command	暂时退出 Vim 编辑器到命令模式下，执行 command 的显示结果，如使用"：!ls/home"即可在 Vim 编辑器中查看/home 下以 ls 输出的文件信息
：set nu	显示行号，设定之后，会在每一行的前面显示该行的行号
：set nonu	与：set nu 相反，即取消行号

在末行模式中可以进行查找与替换操作，其命令格式如下。

:[range]   s/pattern/string/[c,e,g,i]

查找与替换操作各参数及功能说明如表 2.7 所示。

表 2.7 查找与替换操作各参数及功能说明

参数	功能说明
range	指的是范围。"1,5"指从第 1 行至第 5 行，"1,$"指从第 1 行至最后一行，即整篇文章
s（Search）	表示查找搜索
pattern	要被替换的字符串
string	用 string 替换 pattern 内容
c（Confirm）	每次替换前会询问
e（Error）	不显示 error
g（Globe）	不询问，将进行整行替换
i（Ignore）	不区分大小写

在命令模式中输入"/"或"?"字符，也可以进入末行模式，在末行模式中可以进行查找操作，其命令格式如下。

/word

或者

? word

查找操作各参数及功能说明如表 2.8 所示。

表 2.8 查找操作各参数及功能说明

参数	功能说明
/word	从光标位置之下寻找一个名为 word 的字符串,例如,要在文件内查找"welcome"这个字符串,则可输入/welcome
? word	从光标位置之上寻找一个名为 word 的字符串
n	重复前一个查找的动作,例如,执行/welcome 向下查找 welcome 这个字符串,则按下 n 键后,会向下继续查找下一个名为 welcome 的字符串,如果是执行? welcome,那么按下 n 键会向上继续查找名为 welcome 的字符串
N	与 n 刚好相反,即反向进行前一个查找动作,例如,执行/welcome 操作后,按 N 键表示向上查找 welcome

2. 系统克隆

人们经常用虚拟机做各种实验,初学者免不了误操作导致系统崩溃、无法启动,或者在做集群的时候,通常需要使用多台服务器进行测试,如搭建 MySQL 服务、Redis 服务、Tomcat、Nginx 等。搭建一台服务器费时费力,一旦系统崩溃、无法启动,需要重新安装操作系统或部署多台服务器的时候,将会浪费很多时间。那么如何进行操作呢?系统克隆可以很好地解决这个问题。

在虚拟机安装好原始的操作系统后,进行系统克隆,多克隆出几份并备用,方便日后多台机器进行实验测试,这样就可以避免重新安装操作系统,方便快捷。

(1)打开 VMware 虚拟机主窗口,关闭虚拟机中的操作系统,选择要克隆的操作系统,选择"虚拟机"→"管理"→"克隆"选项,如图 2.1 所示。

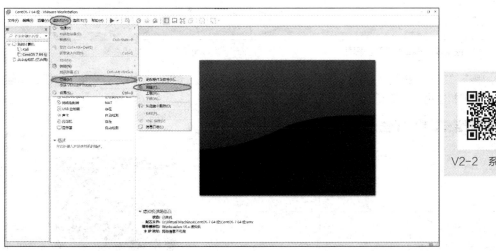

V2-2 系统克隆

图 2.1 系统克隆

(2)弹出克隆虚拟机向导界面,如图 2.2 所示。单击"下一步"按钮,弹出选择克隆源界面,如图 2.3 所示,可以选中"虚拟机中的当前状态"或"现有快照(仅限关闭的虚拟机)"单选按钮。

(3)单击"下一步"按钮,弹出选择克隆类型界面,如图 2.4 所示。选择克隆方法,可以选中"创建链接克隆"单选按钮,也可以选中"创建完整克隆"单选按钮。

(4)单击"下一步"按钮,新虚拟机名称界面,如图 2.5 所示,为虚拟机命名并进行安装位置的设置。

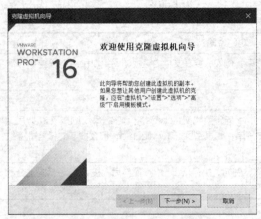

图 2.2　克隆虚拟机向导界面　　　　　图 2.3　选择克隆源界面

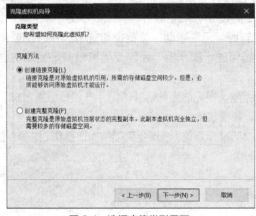

图 2.4　选择克隆类型界面　　　　　图 2.5　新虚拟机名称界面

（5）单击"完成"按钮，弹出正在克隆虚拟机界面，如图 2.6 所示。单击"关闭"按钮，返回 VMware 虚拟机主窗口，系统克隆完成，如图 2.7 所示。

图 2.6　正在克隆虚拟机界面　　　　　图 2.7　系统克隆完成

3. 快照管理

VMware 快照是 VMware Workstation 的一个特色功能，当用户创建一个虚拟机快照时，它会创建一个特定的文件 delta。delta 文件是在 VMware 虚拟机磁盘格式（Virtual Machine Disk Format，VMDK）文件上的变更位图，因此，它不能比 VMDK 还大。每为虚拟机创建一个快照,都会创建一个 delta 文件,当快照被删除或在快照管理中被恢复时,文件将自动被删除。

可以把虚拟机某个时间点的内存、磁盘文件等的状态保存为一个镜像文件。通过这个镜像文件,用户可以在以后的任何时间来恢复虚拟机创建快照时的状态。日后系统出现问题时,可以从快照中进行恢复。

V2-3　系统快照

(1)打开 VMware 虚拟机主窗口,启动虚拟机中的系统,选择要快照保存备份的内容,选择"虚拟机"→"快照"→"拍摄快照"选项,如图 2.8 所示。命名快照名称,如图 2.9 所示。

(2)单击"拍摄快照"按钮,返回 VMware 虚拟机主窗口,拍摄快照完成,如图 2.10 所示。

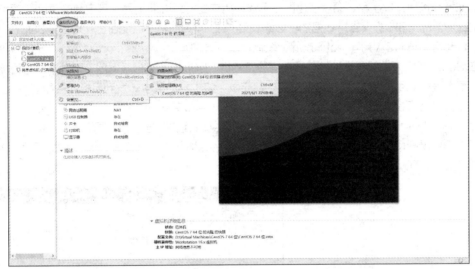

图 2.8　拍摄快照

图 2.9　命名快照名称　　　　　　　　图 2.10　拍摄快照完成

4. SecureCRT 远程连接管理 Linux 操作系统

SecureCRT(Combined Rlogin and Telnet,CRT)和 SecureFX(FTP、SFTP 和 FTP over SSH2,FX)都是由 VanDyke Software 公司出品的安全外壳(Secure Shell,SSH)传输工具。SecureCRT 可以进行远程连接,SecureFX 可以进行远程可视化文件传输。

SecureCRT 是一款支持 SSH（SSH1 和 SSH2）的终端仿真程序，简单地说，其是 Windows 操作系统中登录 UNIX 或 Linux 服务器主机的软件。

SecureCRT 支持 SSH，同时支持 Telnet 和 Rlogin 协议。SecureCRT 是一款用于连接运行 Windows、UNIX 和虚拟内存系统（Virtual Memory System，VMS）等的理想工具；通过使用内含的向量通信处理器（Vector Communication Processor，VCP），命令行程序可以进行加密文件的传输；有 CRTTelnet 客户机的所有特点，包括自动注册、对不同主机保持不同的特性、打印功能、颜色设置、可变屏幕尺寸、用户定义的键位图和优良的 VT100、VT102、VT220，以及全新微小的整合（All New Small Integration，ANSI）竞争，能在命令行中运行或在浏览器中运行。其他特点包括文本手稿、易于使用的工具条、用户的键位图编辑器、可定制的 ANSI 颜色等。SecureCRT 的 SSH 协议支持数据加密标准（Data Encryption Standard，DES）、3DES、RC4 密码，以及密码与 RSA（Rivest Shamir Adleman）鉴别。

V2-4 远程连接管理 Linux 系统

在 SecureCRT 中配置本地端口转发，涉及本机、跳板机、目标服务器，因为本机与目标服务器不能直接进行 ping 操作，所以需要配置端口转发，将本机的请求转发到目标服务器。

（1）为了方便操作，使用 SecureCRT 连接 Linux 服务器，选择相应的虚拟机操作系统。在 VMware 虚拟机主窗口中，选择"编辑"→"虚拟网络编辑器"选项，如图 2.11 所示。

（2）在"虚拟网络编辑器"对话框中，选择"VMnet8"选项，设置 NAT 模式的子网 IP 地址为 192.168.100.0，如图 2.12 所示。

图 2.11 选择"虚拟网络编辑器"选项

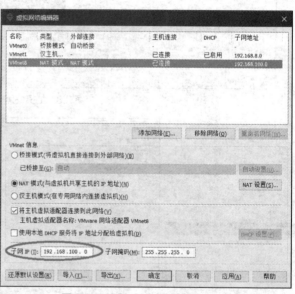

图 2.12 设置 NAT 模式的子网 IP 地址

（3）在"虚拟网络编辑器"对话框中，单击"NAT 设置"按钮，弹出"NAT 设置"对话框，设置网关 IP 地址，如图 2.13 所示。

（4）选择"控制面板"→"网络和 Internet"→"网络连接"选项，查看 VMware Network Adapter VMnet8 连接，如图 2.14 所示。

（5）选择 VMnet8 的 IP 地址，如图 2.15 所示。

（6）进入 Linux 操作系统桌面，单击桌面右上角的"启动"按钮 ⏻，选择"有线连接 已关闭"选项，设置网络有线连接，如图 2.16 所示。

图 2.13 设置网关 IP 地址

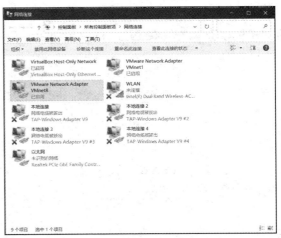

图 2.14 查看 VMware Network Adapter VMnet8 连接

图 2.15 选择 VMnet8 的 IP 地址

图 2.16 设置网络有线连接

(7) 选择"有线设置"选项,打开"设置"窗口,如图 2.17 所示。

图 2.17 "设置"窗口

（8）在"设置"窗口中单击"有线连接" 按钮，选择"IPv4"选项卡，设置IPv4信息，如IP地址、子网掩码、网关、域名服务（Domain Name Service，DNS）相关信息，如图2.18所示。

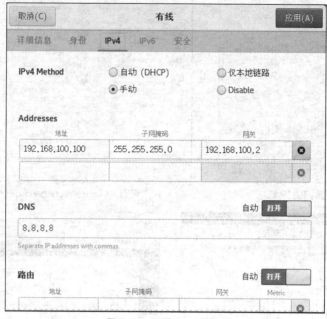

图2.18　设置IPv4信息

（9）设置完成后，单击"应用"按钮，返回"设置"窗口，单击"关闭"按钮，使按钮变为打开状态。单击"有线连接" 按钮，查看网络配置详细信息，如图2.19所示。

图2.19　查看网络配置详细信息

（10）在Linux操作系统中，使用Firefox浏览器访问网站，如图2.20所示。

图 2.20　使用 Firefox 浏览器访问网站

（11）使用"Windows+R"组合键，打开"运行"对话框，输入命令"cmd"，单击"确定"按钮，如图 2.21 所示。

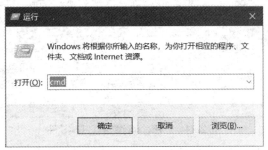

图 2.21　"运行"对话框

（12）使用 ping 命令访问网络主机 192.168.100.100，测试网络连通性，如图 2.22 所示。

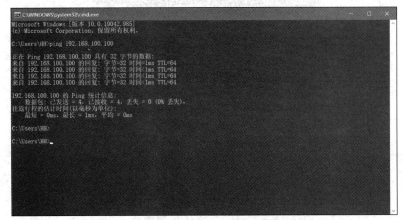

图 2.22　访问网络主机

（13）下载并安装 SecureCRT 工具软件，如图 2.23 所示。

图 2.23　安装 SecureCRT 工具软件

（14）打开 SecureCRT 工具软件，单击工具栏中的图标，如图 2.24 所示。

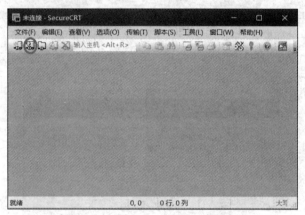

图 2.24　打开 SecureCRT 工具软件

（15）打开"快速连接"对话框，输入主机名为 192.168.100.100，用户名为 root，进行连接，如图 2.25 所示。

图 2.25　SecureCRT 的"快速连接"对话框

（16）打开"新建主机密钥"对话框，提示相关信息，如图 2.26 所示。

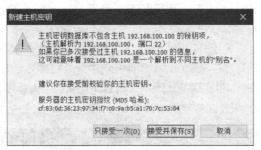

图 2.26 "新建主机密钥"对话框

（17）单击"接受并保存"按钮，打开"输入安全外壳密码"对话框，输入用户名和密码，如图 2.27 所示。

图 2.27 SecureCRT 的"输入安全外壳密码"对话框

（18）单击"确定"按钮，出现图 2.28 所示结果，表示已经成功连接网络主机 192.168.100.100。

图 2.28 成功连接网络主机

5. SecureFX 远程连接文件传送配置

SecureFX 支持 3 种文件传送协议：文件传送协议（File Transfer Protocol，FTP）、安全文件传送协议（Secure File Transfer Protocol，SFTP）和 FTP over SSH2。无论用户连接的是哪种操作系统的服务器，它都能提供安全的传送服务。它主要用于 Linux 操作系统，如 Red Hat、Ubuntu 的客户端文件传送，用户可以选择利用 SFTP 通过加密的 SSH2 实现安全传送，也可以利用 FTP 进行标准传送。该客户端具有 Explorer 风格的界面，易于使用，同时提供强大的自动化功能，可以实现自动化的安全文件传送。

SecureFX 可以更加有效地实现文件的安全传送，用户可以使用其新的拖放功能直接将文件拖放至 Windows Explorer 或其他程序中，也可以充分利用 SecureFX 的自动化特性，实现无须人为干扰的文件自动传送。新版 SecureFX 采用了一个密码库，符合 FIPS 140-2 加密要求，改进了 X.509 证书的认证能力，可以轻松开启多个会话，提高了 SSH 代理的功能。

总的来说，SecureCRT 是在 Windows 操作系统中登录 UNIX 或 Linux 服务器主机的软件，SecureFX 是一款 FTP 软件，可实现 Windows 和 UNIX 或 Linux 操作系统的文件互动。

（1）下载并安装 SecureFX 工具软件，如图 2.29 所示。

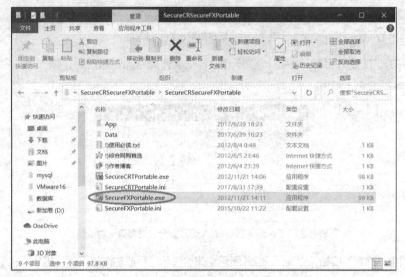

图 2.29 安装 SecureFX 工具软件

（2）打开 SecureFX 工具软件，单击工具栏中的图标 ，如图 2.30 所示。

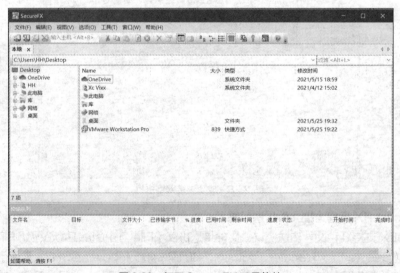

图 2.30 打开 SecureFX 工具软件

（3）打开"快速连接"对话框，输入主机名为 192.168.100.100，用户名为 root，进行连接，如图 2.31 所示。

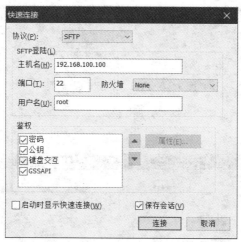

图 2.31 SecureFX 的"快速连接"对话框

(4)在"输入安全外壳密码"对话框中,输入用户名和密码,进行登录,如图 2.32 所示。

图 2.32 SecureFX 的"输入安全外壳密码"对话框

(5)单击"确定"按钮,弹出 SecureFX 主界面,中间部分显示乱码,如图 2.33 所示。

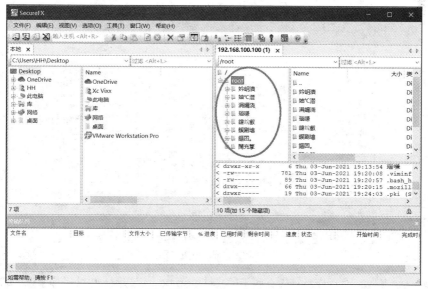

图 2.33 SecureFX 主界面

(6)在 SecureFX 主界面中,选择"选项"→"会话选项"选项,如图 2.34 所示。

图2.34 选择"会话选项"选项

（7）在"会话选项"对话框中，选择"外观"选项，在"字符编码"下拉列表中选择"UTF-8"选项，如图2.35所示。

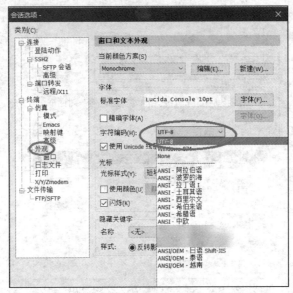

图2.35 设置会话选项

（8）配置完成后，再次显示/boot目录配置结果，如图2.36所示。

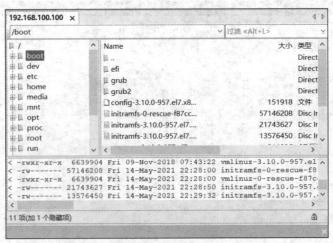

图2.36 显示配置结果

（9）将Windows 10操作系统中F盘下的文件abc.txt，传送到Linux操作系统中的/mnt/aaa目录下。在Linux操作系统中的/mnt/目录下，新建aaa文件夹。选中aaa文件夹，同时选择F盘下的文件abc.txt，并将其拖放到传送队列中，如图2.37所示。

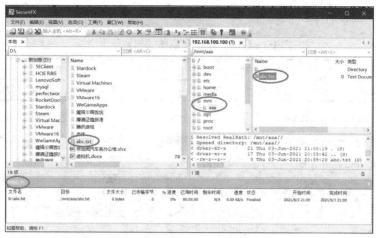

图 2.37　使用 SecureFX 传送文件

（10）使用 ls 命令，查看网络主机 192.168.100.100 目录/mnt/aaa 的传送结果，如图 2.38 所示。

图 2.38　查看网络主机 192.168.100.100 目录/mnt/aaa 的传送结果

2.2.2　云计算管理平台部署需求与规划

云计算管理平台部署需要进行统一规划与设计，如主机名及 IP 地址规划，以及安装前的环境准备工作。具体步骤如下。

1. 主机名及 IP 地址规划

本项目使用安装了 CentOS 7.6 操作系统的主机，安装部署 OpenStack 环境对硬件设备的最低配置要求如表 2.9 所示。

表 2.9　安装部署 OpenStack 环境对硬件设备的最低配置要求

硬件需求	详细信息
CPU	支持 Intel 64 位或 AMD 64 位，并启用了 Intel-TV 或 AMD-TV 硬件虚拟化支持的 64 位 x86 处理器，逻辑 CPU 个数为 4
内存	8GB
磁盘空间	30GB
网络	1 块传输速率为 1Gbit/s 的网卡

OpenStack 具体的部署环境情况如表 2.10 所示。

表 2.10　OpenStack 具体的部署环境情况

主机名	IP 地址	角色
openstack	192.168.100.100/24	安装所有 OpenStack 的组件及需要的环境

2. 安装前的环境准备工作

在正式部署 OpenStack 之前,需要准备如下环境。

(1)修改主机名,配置静态 IP 地址及网关、DNS 参数,并测试网络连通性。

[root@localhost ~]# hostnamectl set-hostname openstack
[root@localhost ~]# bash
[root@openstack ~]# vim /etc/sysconfig/network-scripts/ifcfg-ens33
修改选项:
BOOTPROTO=dhcp--->static　　　　　　　//配置 DHCP,配置为静态
ONBOOT=no--->yes　　　　　　　　　　//是否激活网卡,配置为激活状态
增加选项:
IPADDR=192.168.100.100　　　　　　　　//配置 IP 地址
PREFIX=24 或 NETMASK=255.255.255.0　　//配置子网掩码
GATEWAY=192.168.100.2　　　　　　　　//配置网关
DNS1=8.8.8.8　　　　　　　　　　　　　//配置 DNS 的 IP 地址
[root@openstack ~]# systemctl restart network　　//重启网络服务

配置完成后,使用 cat 命令,查看配置的静态 IP 地址及网关、DNS 参数,如图 2.39 所示。

图 2.39　查看配置的静态 IP 地址及网关、DNS 参数

使用 ping 命令,测试外部网络连通性,这里以网易网为例进行测试,如图 2.40 所示。

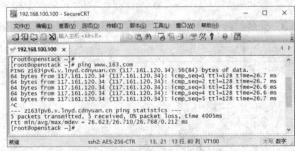

图 2.40　测试外部网络连通性

（2）取消防火墙开机启动功能，执行操作如下。

```
[root@openstack ~]# systemctl disable firewalld
[root@openstack ~]# systemctl stop firewalld
```

（3）取消 NetworkManager 开机启动功能，执行操作如下。

```
[root@openstack ~]# systemctl disable NetworkManager
[root@openstack ~]# systemctl stop NetworkManager
```

（4）关闭 SELinux 开机启动功能，执行操作如下。

```
[root@openstack ~]# cat /etc/sysconfig/selinux
# This file controls the state of SELinux on the system.
# SELINUX= can take one of these three values:
#     enforcing - SELinux security policy is enforced.
#     permissive - SELinux prints warnings instead of enforcing.
#     disabled - No SELinux policy is loaded.
SELINUX=disabled
# SELINUXTYPE= can take one of three values:
#     targeted - Targeted processes are protected,
#     minimum - Modification of targeted policy. Only selected processes are protected.
#     mls - Multi Level Security protection.
SELINUXTYPE=targeted
[root@openstack ~]#
```

（5）启动网络服务功能，执行操作如下。

```
[root@openstack ~]# systemctl enable network
[root@openstack ~]# systemctl start network
```

（6）重启主机系统，执行操作如下。

```
[root@openstack ~]# reboot
```

2.3 项目实施

2.3.1 使用 Packstack 一键部署 OpenStack 云计算管理平台

完成环境准备之后，接下来通过 Packstack 部署 OpenStack。Packstack 是能够自动部署 OpenStack 的工具。通过它管理员可完成 OpenStack 的自动部署。

为了完成这一目标，需要先通过 YUM 源安装 Packstack 工具，再利用 Packstack 工具一键部署 OpenStack。

1. 配置 YUM 源

最小化安装 CentOS 7.6 之后，系统会默认提供 CentOS 的官方 YUM 源。官方 YUM 源中包含了用于部署 OpenStack 各种版本的安装源，本项目选择安装 Stein 版本。

（1）备份 YUM 源文件。

```
[root@localhost ~]# mkdir  /mnt/databackup  -p         //创建备份目录
[root@localhost ~]# mv  /etc/yum.repos.d/*  /mnt/databackup/
[root@localhost ~]# ls  /etc/yum.repos.d/
[root@localhost ~]# ls  /mnt/databackup/
CentOS-Base.repo      CentOS-Debuginfo.repo    CentOS-Media.repo      CentOS-Vault.repo
CentOS-CR.repo        CentOS-fasttrack.repo    CentOS-Sources.repo
```

[root@localhost ~]#

（2）使用 curl 命令下载新的阿里镜像源 CentOS-Base.repo 到/etc/yum.repos.d/目录下。

[root@localhost ~]# curl -o /etc/yum.repos.d/CentOS-Base.repo http://mirrors.aliyun.com/repo/Centos-7.repo

```
  % Total    % Received % Xferd  Average Speed   Time    Time     Time  Current
                                 Dload  Upload   Total   Spent    Left  Speed
100  2523  100  2523    0     0  13572      0 --:--:-- --:--:-- --:--:-- 13637
```
[root@localhost ~]# ls /etc/yum.repos.d/
CentOS-Base.repo
[root@localhost ~]#

或使用 wget 命令下载新的阿里镜像源 CentOS-Base.repo 到/etc/yum.repos.d/目录下。

[root@localhost ~]# cd /etc/yum.repos.d/
[root@localhost yum.repos.d]# rm CentOS-Base.repo
rm：是否删除普通文件 "CentOS-Base.repo"? Y
[root@localhost yum.repos.d]# ls
[root@localhost yum.repos.d]# wget http://mirrors.aliyun.com/repo/Centos-7.repo
--2020-10-08 22:10:06-- http://mirrors.aliyun.com/repo/Centos-7.repo
正在解析主机 mirrors.aliyun.com (mirrors.aliyun.com)... 120.201.245.131, 120.201.245.129, 120.201.245.132, ...
正在连接 mirrors.aliyun.com (mirrors.aliyun.com)|120.201.245.131|:80... 已连接。
已发出 HTTP 请求，正在等待回应... 200 OK
长度：2523 (2.5K) [application/octet-stream]
正在保存至："Centos-7.repo"

100%[===>] 2,523   --.-K/s 用时 0s

2020-10-08 22:10:06 (189 MB/s) - 已保存 "Centos-7.repo" [2523/2523]

[root@localhost yum.repos.d]# mv Centos-7.repo CentOS-Base.repo
[root@localhost yum.repos.d]# ls
CentOS-Base.repo
[root@localhost yum.repos.d]#

（3）查看阿里镜像源 CentOS-Base.repo 文件的内容。

[root@localhost yum.repos.d]# cat CentOS-Base.repo
CentOS-Base.repo
#
The mirror system uses the connecting IP address of the client and the
update status of each mirror to pick mirrors that are updated to and
geographically close to the client. You should use this for CentOS updates
unless you are manually picking other mirrors.
#
If the mirrorlist= does not work for you, as a fall back you can try the
remarked out baseurl= line instead.
#
[base]
name=CentOS-$releasever - Base - mirrors.aliyun.com
failovermethod=priority
baseurl=http://mirrors.aliyun.com/centos/$releasever/os/$basearch/
        http://mirrors.aliyuncs.com/centos/$releasever/os/$basearch/

```
             http://mirrors.cloud.aliyuncs.com/centos/$releasever/os/$basearch/
gpgcheck=1
gpgkey=http://mirrors.aliyun.com/centos/RPM-GPG-KEY-CentOS-7
 #released updates
[updates]
name=CentOS-$releasever - Updates - mirrors.aliyun.com
failovermethod=priority
baseurl=http://mirrors.aliyun.com/centos/$releasever/updates/$basearch/
             http://mirrors.aliyuncs.com/centos/$releasever/updates/$basearch/
             http://mirrors.cloud.aliyuncs.com/centos/$releasever/updates/$basearch/
gpgcheck=1
gpgkey=http://mirrors.aliyun.com/centos/RPM-GPG-KEY-CentOS-7
 #additional packages that may be useful
[extras]
name=CentOS-$releasever - Extras - mirrors.aliyun.com
failovermethod=priority
baseurl=http://mirrors.aliyun.com/centos/$releasever/extras/$basearch/
             http://mirrors.aliyuncs.com/centos/$releasever/extras/$basearch/
             http://mirrors.cloud.aliyuncs.com/centos/$releasever/extras/$basearch/
gpgcheck=1
gpgkey=http://mirrors.aliyun.com/centos/RPM-GPG-KEY-CentOS-7
 #additional packages that extend functionality of existing packages
[centosplus]
name=CentOS-$releasever - Plus - mirrors.aliyun.com
failovermethod=priority
baseurl=http://mirrors.aliyun.com/centos/$releasever/centosplus/$basearch/
             http://mirrors.aliyuncs.com/centos/$releasever/centosplus/$basearch/
             http://mirrors.cloud.aliyuncs.com/centos/$releasever/centosplus/$basearch/
gpgcheck=1
enabled=0
gpgkey=http://mirrors.aliyun.com/centos/RPM-GPG-KEY-CentOS-7
 #contrib - packages by CentOS Users
[contrib]
name=CentOS-$releasever - Contrib - mirrors.aliyun.com
failovermethod=priority
baseurl=http://mirrors.aliyun.com/centos/$releasever/contrib/$basearch/
             http://mirrors.aliyuncs.com/centos/$releasever/contrib/$basearch/
             http://mirrors.cloud.aliyuncs.com/centos/$releasever/contrib/$basearch/
gpgcheck=1
enabled=0
gpgkey=http://mirrors.aliyun.com/centos/RPM-GPG-KEY-CentOS-7
[root@localhost yum.repos.d]# cd ~
[root@localhost ~]#
```

2. 配置 Shell 脚本

自动禁用 firewalld、禁用 NetworkManager 并配置网络。

```
[root@localhost ~]# vim  forbidden.sh
#!/bin/bash
```

```
hostnamectl set-hostname openstack
bash
systemctl disable firewalld
systemctl stop firewalld
systemctl disable NetworkManager
systemctl stop NetworkManager
systemctl enable network
systemctl start network
[root@localhost ~]# . forbidden.sh       //执行脚本
[root@openstack ~]#
```

3. 设置 OpenStack 存储库

```
[root@openstack ~]# yum clean all
已加载插件：fastestmirror, langpacks
正在清理软件源： base extras updates
Cleaning up list of fastest mirrors
[root@openstack ~]# yum repolist all
已加载插件：fastestmirror, langpacks
Determining fastest mirrors
 * base: mirrors.aliyun.com
 * extras: mirrors.aliyun.com
 * updates: mirrors.aliyun.com
base                                                         | 3.6 kB  00:00:00
extras                                                       | 2.9 kB  00:00:00
updates                                                      | 2.9 kB  00:00:00
(1/4): extras/7/x86_64/primary_db                            | 206 kB  00:00:00
(2/4): base/7/x86_64/group_gz                                | 153 kB  00:00:00
(3/4): updates/7/x86_64/primary_db                           | 4.5 MB  00:00:00
base/7/x86_64/primary_db        FAILED
http://mirrors.aliyuncs.com/centos/7/os/x86_64/repodata/f09552edffa70f49f553e411c2282fbccff
fbeafa21e81e32622b103038b8bae-primary.sqlite.bz2: [Errno 14] curl#7 - "Failed connect to
mirrors.aliyuncs.com:80; Connection refused"
正在尝试其他镜像。
(4/4): base/7/x86_64/primary_db                              | 6.1 MB  00:00:00
源标识                          源名称                                   状态
base/7/x86_64          CentOS-7 - Base - mirrors.aliyun.com       启用: 10,070
centosplus/7/x86_64    CentOS-7 - Plus - mirrors.aliyun.com       禁用
contrib/7/x86_64       CentOS-7 - Contrib - mirrors.aliyun.com    禁用
extras/7/x86_64        CentOS-7 - Extras - mirrors.aliyun.com     启用:    413
updates/7/x86_64       CentOS-7 - Updates - mirrors.aliyun.com    启用:  1,134
repolist: 11,617
[root@openstack ~]#
```

在 CentOS 7.6 中，Extras 资源库提供了启用 OpenStack 资源库的红帽软件包管理器（Red Hat Package Manager，RPM）。默认情况下，Extras 在 CentOS 7.6 中已经启用了，因此只需安装 RPM 即可设置 OpenStack 存储库。

执行以下命令，可以设置 OpenStack 所需要的软件库，本例使用的 OpenStack 的版本为 Stein。

```
[root@openstack ~]# yum install -y centos-release-openstack-stein
已加载插件：fastestmirror, langpacks
Loading mirror speeds from cached hostfile
 * base: mirrors.aliyun.com
 * extras: mirrors.aliyun.com
 * updates: mirrors.aliyun.com
正在解决依赖关系
--> 正在检查事务
---> 软件包 centos-release-openstack-stein.noarch.0.1-1.el7.centos 将被 安装
--> 正在处理依赖关系 centos-release-qemu-ev，它被软件包 centos-release-openstack-stein-1-1.el7.centos.noarch 需要
--> 正在处理依赖关系 centos-release-ceph-nautilus，它被软件包 centos-release-openstack-stein-1-1.el7.centos.noarch 需要
--> 正在检查事务
---> 软件包 centos-release-ceph-nautilus.noarch.0.1.2-2.el7.centos 将被 安装
--> 正在处理依赖关系 centos-release-storage-common，它被软件包 centos-release-ceph-nautilus-1.2-2.el7.centos.noarch 需要
--> 正在处理依赖关系 centos-release-nfs-ganesha28，它被软件包 centos-release-ceph-nautilus-1.2-2.el7.centos.noarch 需要
---> 软件包 centos-release-qemu-ev.noarch.0.1.0-4.el7.centos 将被 安装
--> 正在处理依赖关系 centos-release-virt-common，它被软件包 centos-release-qemu-ev-1.0-4.el7.centos.noarch 需要
--> 正在检查事务
---> 软件包 centos-release-nfs-ganesha28.noarch.0.1.0-3.el7.centos 将被 安装
---> 软件包 centos-release-storage-common.noarch.0.2-2.el7.centos 将被 安装
---> 软件包 centos-release-virt-common.noarch.0.1-1.el7.centos 将被 安装
--> 解决依赖关系完成
依赖关系解决

================================================================================
 Package                            架构      版本              源       大小
================================================================================
正在安装:
 centos-release-openstack-stein     noarch    1-1.el7.centos    extras   5.3 k
为依赖而安装:
 centos-release-ceph-nautilus       noarch    1.2-2.el7.centos  extras   5.1 k
 centos-release-nfs-ganesha28       noarch    1.0-3.el7.centos  extras   4.3 k
 centos-release-qemu-ev             noarch    1.0-4.el7.centos  extras   11 k
 centos-release-storage-common      noarch    2-2.el7.centos    extras   5.1 k
 centos-release-virt-common         noarch    1-1.el7.centos    extras   4.5 k
事务概要
================================================================================
安装  1 软件包 (+5 依赖软件包)
总下载量：35 k
安装大小：25 k
Downloading packages:
```

警告：/var/cache/yum/x86_64/7/extras/packages/centos-release-ceph-nautilus-1.2-2.el7.centos.noarch.rpm: 头V3 RSA/SHA256 Signature, 密钥 ID f4a80eb5: NOKEY
centos-release-ceph-nautilus-1.2-2.el7.centos.noarch.rpm 的公钥尚未安装
(1/6): centos-release-ceph-nautilus-1.2-2.el7.centos.noarch.rpm | 5.1 kB 00:00:00
(2/6): centos-release-nfs-ganesha28-1.0-3.el7.centos.noarch.rpm | 4.3 kB 00:00:00
(3/6): centos-release-qemu-ev-1.0-4.el7.centos.noarch.rpm | 11 kB 00:00:00
(4/6): centos-release-openstack-stein-1-1.el7.centos.noarch.rpm | 5.3 kB 00:00:00
(5/6): centos-release-storage-common-2-2.el7.centos.noarch.rpm | 5.1 kB 00:00:00
(6/6): centos-release-virt-common-1-1.el7.centos.noarch.rpm | 4.5 kB 00:00:00

总计   84 kB/s | 35 kB 00:00:00
从 http://mirrors.aliyun.com/centos/RPM-GPG-KEY-CentOS-7 检索密钥
导入 GPG key 0xF4A80EB5:
 用户 ID   : "CentOS-7 Key (CentOS 7 Official Signing Key) <security@centos.org>"
 指纹   : 6341 ab27 53d7 8a78 a7c2 7bb1 24c6 a8a7 f4a8 0eb5
 来自   : http://mirrors.aliyun.com/centos/RPM-GPG-KEY-CentOS-7
Running transaction check
Running transaction test
Transaction test succeeded
Running transaction
  正在安装   : centos-release-storage-common-2-2.el7.centos.noarch   1/6
  正在安装   : centos-release-nfs-ganesha28-1.0-3.el7.centos.noarch   2/6
  正在安装   : centos-release-ceph-nautilus-1.2-2.el7.centos.noarch   3/6
  正在安装   : centos-release-virt-common-1-1.el7.centos.noarch   4/6
  正在安装   : centos-release-qemu-ev-1.0-4.el7.centos.noarch   5/6
  正在安装   : centos-release-openstack-stein-1-1.el7.centos.noarch   6/6
  验证中   : centos-release-openstack-stein-1-1.el7.centos.noarch   1/6
  验证中   : centos-release-virt-common-1-1.el7.centos.noarch   2/6
  验证中   : centos-release-ceph-nautilus-1.2-2.el7.centos.noarch   3/6
  验证中   : centos-release-nfs-ganesha28-1.0-3.el7.centos.noarch   4/6
  验证中   : centos-release-storage-common-2-2.el7.centos.noarch   5/6
  验证中   : centos-release-qemu-ev-1.0-4.el7.centos.noarch   6/6
已安装:
  centos-release-openstack-stein.noarch 0:1-1.el7.centos
作为依赖被安装:
  centos-release-ceph-nautilus.noarch   0:1.2-2.el7.centos
centos-release-nfs-ganesha28.noarch 0:1.0-3.el7.centos
  centos-release-qemu-ev.noarch   0:1.0-4.el7.centos
centos-release-storage-common.noarch 0:2-2.el7.centos

```
  centos-release-virt-common.noarch 0:1-1.el7.centos
完毕！
[root@openstack ~]#
```

4．安装 Packstack 工具

Packstack 是一款 Windows 操作系统中的远程桌面管理软件，可让用户在同一个窗口中浏览多个远程桌面的信息，方便 Windows 操作系统远程管理。其用于取代手动设置 OpenStack。Packstack 基于 Puppet 工具，通过 Puppet 部署 Packstack 各组件。Puppet 是一种 Linux 和 Windows 平台的集中配置管理系统，使用 Puppet 自有的描述语言，可管理配置文件、用户、任务、软件包、系统服务等。

执行以下命令，安装 openstack-packstack 及其依赖包。

```
[root@openstack ~]# yum install -y openstack-packstack
已加载插件：fastestmirror, langpacks
Loading mirror speeds from cached hostfile
 * base: mirrors.aliyun.com
 * centos-ceph-nautilus: mirror.bit.edu.cn
 * centos-nfs-ganesha28: mirrors.neusoft.edu.cn
 * centos-openstack-stein: mirrors.neusoft.edu.cn
 * centos-qemu-ev: mirrors.neusoft.edu.cn
 * extras: mirrors.aliyun.com
 * updates: mirrors.aliyun.com
……
  puppet-staging.noarch 0:1.0.4-1.b466d93git.el7
puppet-stdlib.noarch 0:5.2.0-1.b0dd4c1git.el7
  puppet-swift.noarch 0:14.4.0-1.el7                              puppet-sysctl.noarch
0:0.0.12-2.a3d160dgit.el7
  puppet-tempest.noarch 0:14.4.0-1.el7
puppet-trove.noarch 0:14.4.0-1.el7
  puppet-vcsrepo.noarch 0:2.4.0-1.a2ea398git.el7
puppet-vswitch.noarch 0:10.4.0-1.el7
  puppet-xinetd.noarch 0:3.2.0-1.98b4231git.el7
python-docutils.noarch 0:0.11-0.3.20130715svn7687.el7
  python2-asn1crypto.noarch 0:0.23.0-2.el7
python2-olefile.noarch 0:0.44-1.el7
  rubygem-rdoc.noarch 0:4.0.0-36.el7        rubygem-rgen.noarch 0:0.6.6-2.el7
  rubygems.noarch 0:2.0.14.1-36.el7         yaml-cpp.x86_64 0:0.5.1-6.el7
作为依赖被升级：
  libselinux.x86_64 0:2.5-15.el7            libselinux-python.x86_64 0:2.5-15.el7
libselinux-utils.x86_64 0:2.5-15.el7
  python2-cryptography.x86_64 0:2.5-1.el7
替代：
  python-cffi.x86_64 0:1.6.0-5.el7          python-idna.noarch 0:2.4-1.el7
python-ipaddress.noarch 0:1.0.16-2.el7
  python-netaddr.noarch 0:0.7.5-9.el7       python-six.noarch 0:1.9.0-2.el7
完毕！
[root@openstack ~]#
```

5. 运行 Packstack 一键部署 OpenStack

（1）Packstack 工具的基本语法格式如下。

```
packstack  [选项]  [--help]
```

执行 packstack --help 命令列出选项清单，这里给出部分选项及其说明。

--answer-file=ANSWER_FILE：依据应答文件配置信息以非交互模式运行该工具。

--gen-answer-file=GEN_ANSWER_FILE：产生应答文件模板。

--allinone：所有功能都集中安装在单一主机上。

（2）Packstack 一键部署 OpenStack。

执行以下命令，可以一键部署 OpenStack。

```
[root@openstack ~]# packstack --allinone
Welcome to the Packstack setup utility
The installation log file is available at: /var/tmp/packstack/20201008-230030-3T0RX3/openstack-setup.log
Packstack changed given value  to required value /root/.ssh/id_rsa.pub
Installing:
Clean Up                                              [ DONE ]
Discovering ip protocol version                       [ DONE ]
Setting up ssh keys                                   [ DONE ]
Preparing servers                                     [ DONE ]
Pre installing Puppet and discovering hosts' details  [ DONE ]
Preparing pre-install entries                         [ DONE ]
Setting up CACERT                                     [ DONE ]
Preparing AMQP entries                                [ DONE ]
Preparing MariaDB entries                             [ DONE ]
Fixing Keystone LDAP config parameters to be undef if empty[ DONE ]
Preparing Keystone entries                            [ DONE ]
Preparing Glance entries                              [ DONE ]
Checking if the Cinder server has a cinder-volumes vg[ DONE ]
Preparing Cinder entries                              [ DONE ]
Preparing Nova API entries                            [ DONE ]
Creating ssh keys for Nova migration                  [ DONE ]
Gathering ssh host keys for Nova migration            [ DONE ]
Preparing Nova Compute entries                        [ DONE ]
Preparing Nova Scheduler entries                      [ DONE ]
Preparing Nova VNC Proxy entries                      [ DONE ]
Preparing OpenStack Network-related Nova entries      [ DONE ]
Preparing Nova Common entries                         [ DONE ]
Preparing Neutron LBaaS Agent entries                 [ DONE ]
Preparing Neutron API entries                         [ DONE ]
Preparing Neutron L3 entries                          [ DONE ]
Preparing Neutron L2 Agent entries                    [ DONE ]
Preparing Neutron DHCP Agent entries                  [ DONE ]
Preparing Neutron Metering Agent entries              [ DONE ]
Checking if NetworkManager is enabled and running     [ DONE ]
Preparing OpenStack Client entries                    [ DONE ]
Preparing Horizon entries                             [ DONE ]
```

```
Preparing Swift builder entries                            [ DONE ]
Preparing Swift proxy entries                              [ DONE ]
Preparing Swift storage entries                            [ DONE ]
Preparing Gnocchi entries                                  [ DONE ]
Preparing Redis entries                                    [ DONE ]
Preparing Ceilometer entries                               [ DONE ]
Preparing Aodh entries                                     [ DONE ]
Preparing Puppet manifests                                 [ DONE ]
Copying Puppet modules and manifests                       [ DONE ]
Applying 192.168.100.100_controller.pp
Testing if puppet apply is finished: 192.168.100.100_controller.pp   [ | ]
192.168.100.100_controller.pp:                             [ DONE ]
Applying 192.168.100.100_network.pp
192.168.100.100_network.pp:                                [ DONE ]
Applying 192.168.100.100_compute.pp
192.168.100.100_compute.pp:                                [ DONE ]
Applying Puppet manifests                                  [ DONE ]
Finalizing                                                 [ DONE ]

 **** Installation completed successfully ******
Additional information:
 * Parameter CONFIG_NEUTRON_L2_AGENT: You have choosen OVN neutron backend. Note that this backend does not support LBaaS, VPNaaS or FWaaS services. Geneve will be used as encapsulation method for tenant networks
 * A new answerfile was created in: /root/packstack-answers-20201008-230031.txt
 * Time synchronization installation was skipped. Please note that unsynchronized time on server instances might be problem for some OpenStack components.
 * Warning: NetworkManager is active on 192.168.100.100. OpenStack networking currently does not work on systems that have the Network Manager service enabled.
 * File /root/keystonerc_admin has been created on OpenStack client host 192.168.100.100. To use the command line tools you need to source the file.
 * To access the OpenStack Dashboard browse to http://192.168.100.100/dashboard .
Please, find your login credentials stored in the keystonerc_admin in your home directory.
 * The installation log file is available at: /var/tmp/packstack/20201008-230030-3T0RX3/openstack-setup.log
 * The generated manifests are available at: /var/tmp/packstack/20201008-230030-3T0RX3/manifests
[root@openstack ~]#
```

至此，OpenStack 部署完成。但此时 Linux 虚拟网桥 br-ex 中的 IP 地址是临时的，需要生成配置文件，要使用的命令如下。

```
[root@openstack ~]# cd    /etc/sysconfig/network-scripts/
[root@openstack network-scripts]# cp  ifcfg-ens33   ifcfg-br-ex
[root@openstack network-scripts]# cat   ifcfg-br-ex
TYPE=Ethernet
PROXY_METHOD=none
BROWSER_ONLY=no
BOOTPROTO=static
```

```
DEFROUTE=yes
IPV4_FAILURE_FATAL=no
IPV6INIT=yes
IPV6_AUTOCONF=yes
IPV6_DEFROUTE=yes
IPV6_FAILURE_FATAL=no
IPV6_ADDR_GEN_MODE=stable-privacy
NAME=ens33
UUID=1992e26a-0c1d-4591-bda5-0a2d13c3f5bf
DEVICE=ens33
ONBOOT=yes
IPADDR=192.168.100.100
PREFIX=24
GATEWAY=192.168.100.2
DNS1=8.8.8.8
[root@openstack network-scripts]#
```

至此,已经成功完成 OpenStack 的部署,控制台消息的最后部分给出了环境变量文件和日志文件的位置以及登录 Dashboard 的方法的提示。根据提示,在浏览器的地址栏中输入"http://主机地址/dashboard",可以弹出 OpenStack 的 Horizon Web 界面。在 Horizon Web 界面中,可以与每个 OpenStack 项目 API 进行通信并执行大部分任务。

2.3.2 通过 Dashboard 体验 OpenStack 云计算管理平台功能

Dashboard(仪表板)是 Horizon 项目为 OpenStack 云计算管理平台提供的 Web 访问接口,主要用于 OpenStack 的管理。云管理员和普通用户可以通过 Web 界面管理、控制和使用 OpenStack 云资源及服务。当然,用户也可以直接使用 OpenStack 命令行客户端完成这些任务,这种方式将在后续章节中介绍。

V2-5 Dashboard 界面登录管理 OpenStack 云计算管理平台

1. Dashboard 界面登录管理 OpenStack 云计算管理平台

在客户端的浏览器地址栏中输入 http://192.168.100.100/dashboard,弹出 Dashboard 的登录界面,如图 2.41 所示。

图 2.41 Dashboard 的登录界面

> **注意**
> 如果弹出 500 错误界面,则说明是内部服务器错误,可重新启动服务器以解决问题。

2. Dashboard 的主界面

安装 OpenStack 后,在 root 用户的 root 目录下会生成一个 keystonerc_admin 文件。该文件记录有 Keystone(OpenStack 认证组件)认证的环境变量,包括用户名和登录密码。要使用的命令如下。

```
[root@openstack network-scripts]# cd  ~
[root@openstack ~]# pwd
/root
[root@openstack ~]# ls  -l   keystonerc_admin
-rw-------. 1 root root 375 10 月   8 23:07 keystonerc_admin
[root@openstack ~]# cat   keystonerc_admin                         //显示文件内容
unset OS_SERVICE_TOKEN
    export OS_USERNAME=admin
    export OS_PASSWORD='f4cc880e78b246ee'
    export OS_REGION_NAME=RegionOne
    export OS_AUTH_URL=http://192.168.100.100:5000/v3
    export PS1='[\u@\h \W(keystone_admin)]\$ '
    export OS_PROJECT_NAME=admin
export OS_USER_DOMAIN_NAME=Default
export OS_PROJECT_DOMAIN_NAME=Default
export OS_IDENTITY_API_VERSION=3
    [root@openstack ~]#
[root@openstack ~]#
```

在 Web 控制台中输入用户名和密码并登录后,弹出 Dashboard 默认登录成功管理界面,如图 2.42 所示。如果登录后弹出的是英文界面,则可以进行语言设置。在用户设置窗口中,选择语言为简体中文,如图 2.43 所示。

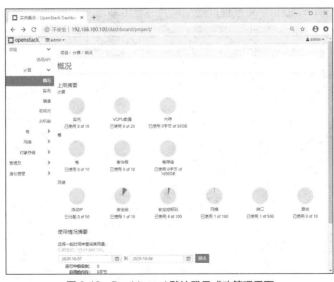

图 2.42 Dashboard 默认登录成功管理界面

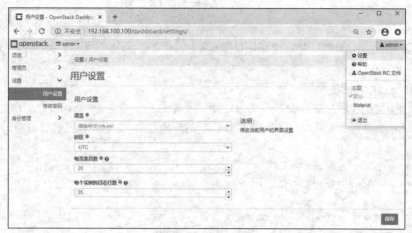

图 2.43 用户设置窗口

课后习题

1. 选择题

（1）通过 Packstack 一键部署 OpenStack 的命令是（　　）。
    A. openstack --allinone　　　　　B. allinone -openstack
    C. packstack --allinone　　　　　D. allinone - packstack

（2）打开 Vim 编辑器，在命令模式下输入（　　），无法进入编辑模式。
    A. a　　　　　B. I　　　　　C. o　　　　　D. d

2. 简答题

（1）简述 Vim 编辑器的 3 种工作模式。
（2）简述如何进行虚拟机克隆与系统快照管理。
（3）简述如何使用 SecureCRT 远程连接及管理 Linux 操作系统。
（4）简述如何使用 SecureFX 远程连接文件并进行传输。
（5）简述如何使用 Packstack 一键部署 OpenStack 云计算管理平台。

项目 3
OpenStack 认证服务

【学习目标】
- 理解Packstack的基本概念。
- 理解Packstack的服务流程。
- 掌握Packstack项目、用户、组、角色管理的不同创建方法。

3.1 项目陈述

在早期的 OpenStack 版本中，用户、消息、API 调用的身份认证都集成在 Nova 项目中，因为加入 OpenStack 的模块越来越多，安全认证所涉及的面越来越广，多种安全认证的处理变得越来越复杂，所以改用一个独立的项目来统一处理不同的认证需求，这个项目就是 Keystone。Keystone 是 OpenStack 身份服务（OpenStack Identity Service）的项目名称，相当于一个别名。Keystone 集成了身份认证、用户授权、用户管理和服务目录等功能，为其他的 Keystone 组件和服务提供了统一的身份服务。Keystone 作为 OpenStack 的一个核心项目，基本上与所有的 OpenStack 项目都相关，当一个 OpenStack 服务收到用户的请求时，首先提交给 Keystone，由它来检查用户是否具有足够的权限来实现其请求的任务。

3.2 必备知识

3.2.1 认证服务基础

Keystone 是 OpenStack 默认使用的身份认证管理系统，也是 OpenStack 中唯一可以提供身份认证的组件。在安装 OpenStack 身份服务之后，其他 OpenStack 服务必须在其中注册才能使用，Keystone 可以跟踪每一个 OpenStack 服务的安装，并在系统网络中定位该服务的位置，身份服务主要用于认证，因此它又称为认证服务。

1. Keystone 的基本概念

Keystone 为每一项 OpenStack 服务都提供了身份服务，而身份服务使用域、项目（租户）、用户和角色等的组合来实现。在讲解 Keystone 之前，有必要介绍以下几个相关的基本概念。

（1）认证

认证（Authentication）是指确认用户身份的过程，又称身份验证。Keystone 认证由用户提供的一组凭证来确认传入请求的有效性。最初，一些凭证是用户名和密码，或者是用户名和 API 密钥。当 Keystone 确认用户凭证有效后，就会发出一个认证令牌（Authentication Token），用户可以在随后的请求中使用这个认证令牌去访问资源中其他的应用。

（2）凭证

凭证（Credentials）又称证书，用于确认用户身份的数据，如用户名、密码和 API 密钥，或认证服务提供的认证令牌。

（3）令牌

令牌（Token）通常指的是一串比特值或者字符串，用来作为访问资源的记号。令牌中含有可访问资源的范围和有效时间，一个令牌可以是一个任意大小的文本，用于与其他 OpenStack 服务共享信息。令牌的有效期是有限的，可以随时被撤回。目前，Keystone 支持基于令牌的认证。

（4）项目

项目（Project）在 OpenStack 的早期版本中被称为租户（Tenant），它是各个服务可以访问资源的集合，是分配和隔离资源或身份的一个容器，也是一种权限组织形式。一个项目可以映射到客户、账户、组织（机构）或租户。OpenStack 用户要访问资源，必须通过一个项目向 Keystone 发出请求，项目是 OpenStack 服务调度的基本单元，其中必须包括相关的用户和角色。

（5）用户

用户（User）是指使用 OpenStack 云服务的个人、系统或服务的账户名称。使用服务的用户可以是人、服务或系统，或使用 OpenStack 相关服务的一个组织。根据不同的安装方式，一个用户可以代表一个客户、账号、组织或项目。OpenStack 各个服务在身份管理体系中都被视为一种系统用户，Keystone 为用户提供认证令牌，使用户在调用 OpenStack 服务时拥有相应的资源使用权限。Keystone 会验证由那些有权限的用户所发出的请求的有效性，用户使用自己的令牌登录和访问资源，用户可以被分配给特定的项目，这样用户就好像包含在该项目中一样，拥有该项目的权限。需要特别指出的是，OpenStack 通过注册相关服务的用户来管理服务，例如，Nova 服务注册 Nova 用户来管理相应的服务，对于管理员来说，需要通过 Keystone 来注册管理用户。

（6）角色

角色（Role）是一个用于定义用户权利和权限的集合，例如，Nova 中的虚拟机、Glance 中的镜像。身份服务向包含一系列角色的用户提供一个令牌，当用户调用服务时，该服务解析用户角色的设置，决定每个角色被授权执行哪些操作或访问哪些资源，通常权限管理是由角色、项目和用户相互配合来实现的。一个项目中往往要包含用户和角色，用户必须依赖于某一项目，而且用户必须以一种角色的身份加入项目，项目正是通过这种方式来实现对项目用户权限规范的绑定的。

（7）组

组（Group）是域所拥有的用户的集合，授予域或项目的组角色可以应用于该组中的所有用户。向组中添加用户，会相应地授予该用户对关联的域或项目的角色和认证；从组中删除用户，也会相应地撤销该用户关联的域或项目的角色和认证。

（8）域

域（Domain）是项目和用户的集合，目的是为身份实体定义管理界限。域可以表示个人、公司或操作人员所拥有的空间，用户可以被授予某个域的管理员角色。域管理员能够在域中创建项目、用户和组，并将角色分配给域中的用户和组。

（9）端点

端点（Endpoint）就是 OpenStack 组件能够访问的网络地址，通常是一个统一资源定位器（Uniform Resource Locator，URL）。端点相当于 OpenStack 服务对外的网络地址列表，每个服务都必须通过端点来检索相应的服务地址。如果需要访问一个服务，则必须知道其端点。端点请求的每个 URL 都对应一个服务实例的访问地址，并且具有 public、private（internal）和 admin 这 3 种权限。public URL 可以被全局访问，private（internal）URL 只能被内部访问，而 admin URL 可以被从常规的访问中分离出来。

（10）客户端

客户端（Client）是一些 OpenStack 服务，包括应用程序接口的命令行接口。例如，用户可以使用 OpenStack service create 和 OpenStack endpoint create 命令，在 OpenStack 安装过程中注册服务。Keystone 的命令行工具可以完成诸如创建用户、角色、服务和端点等绝大多数的 Keystone 管理功能，是常用的命令行接口。

（11）服务

这里的服务（Service）是指计算（Nova）、对象存储（Swift）或镜像（Glance）这样的 OpenStack 服务，它们提供一个或多个端点，供用户访问资源和执行操作。

（12）分区

分区（Region）表示 OpenStack 部署的通用分区。可以为一个分区关联若干个子分区，形成树状层次结构。尽管分区没有地理意义，但是部署时仍可以对分区使用地理名称。

2. Keystone 的主要功能

Keystone 的主要功能如下。

（1）身份认证（Identity Authentication）：令牌的发放和检验。

（2）身份授权（Identity Authorization）：授予用户在一个服务中所拥有的权限。

V3-1　Keystone 的主要功能

（3）用户管理：管理用户账户。

（4）服务目录（Service Catalog）：提供可用服务的 API 端点。

Keystone 在 OpenStack 项目中主要负责以下两个方面的工作。

（1）跟踪用户和监控用户权限。OpenStack 的每个用户和每项服务都必须在 Keystone 中注册，由 Keystone 保存其相关信息。需要身份管理的服务、系统用户都被视为 Keystone 的用户。

（2）为每项 OpenStack 服务提供一个可用的服务目录和相应的 API 端点。OpenStack 身份服务启动之后，一方面，会将 OpenStack 中所有相关的服务置于一个服务列表中，以管理系统能够提供的服务的目录；另一方面，OpenStack 中每个用户会按照各个用户的通用唯一识别码（Universally Unique Identifier，UUID）产生一些 URL，Keystone 受委托管理这些 URL，为需要 API 端点的其他用户提供统一的服务 URL 和 API 调用地址。

3. Keystone 的管理层次结构

在 OpenStack Identity API v3 以前，存在一些需要改进的地方。例如，用户的权限管理以一个用户为单位，需要对每一个用户进行角色分配，并且不存在一种对一组用户进行统一管理的方案，这给系统管理员带了不便，增加了额外的工作量。又如，OpenStack 使用租户来表示一个资源或对象，租户可以包含多个用户，不同租户相互隔离，根据服务运行的需求，租户还可以映射为账户、组织、项目或服务。资源是以租户为单位分配的，不符合现实世界中的层级关系。用户访问一个系统资源时，必须通过一个租户向 Keystone 提出请求。作为 OpenStack 中服务调试的基本单元，租户必须包含相关的用户和角色等信息，并对这两个租户中的用户分别分配角色。OpenStack 租户没有更高层次的单位，无法对多个租户进行统一管理，这就给拥有多租户的企业用户带来了不便。

为了解决上述问题，OpenStack Identity API v3 引入了域和组两个概念，并将租户改称为项目,这样更符合现实世界和云服务的映射关系。

OpenStack Identity API v3 利用域实现了真正的多租户架构,域为项目的高层容器。云服务的客户是域的所有者,它们可以在自己的域中创建多个项目、用户、组和角色。通过引入域,云服务客户可以对其拥有的多个项目进行统一管理,而不必再像之前那样对每一个项目进行单独管理。

V3-2 Keystone 的管理层次结构

组是一组用户的容器,可以向组中添加用户,并直接给组分配角色,这样在这个组中的所有用户就都拥有了该组拥有的角色权限。通过引入组的概念,实现了对用户组的管理,以及同时管理一组用户权限的目的,这与 OpenStack Identity API v2 中直接向用户/项目指定角色不同,它对云服务的管理更加便捷。

域、组、项目、用户和角色之间的关系,即 Keystone 的管理层次结构如图 3.1 所示。一个域中通常包含 3 个项目,可以通过组 Group1 将角色 admin 直接分配给该域,这样组 Group1 中的所有用户将会对域中的所有项目都拥有管理员权限。也可以通过组 Group2 将角色 member 仅分配给项目 Project3,这样组 Group2 中的用户就只拥有对项目 Project3 的相应权限,而不影响其他项目的权限分配。

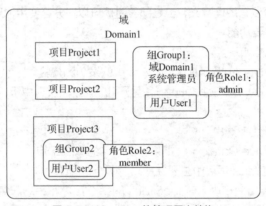

图 3.1 Keystone 的管理层次结构

4. Keystone 的认证服务流程

用户请求云主机的流程涉及认证(Keystone)服务、计算(Nova)服务、镜像(Glance)服务,以及网络(Neutron)服务。在服务流程中,令牌作为流程凭证进行传递。Keystone 的认证服务流程如图 3.2 所示。

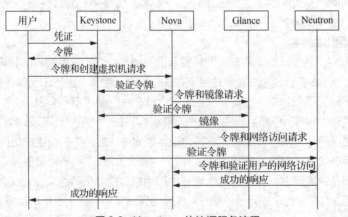

V3-3 Keystone 的认证服务流程

图 3.2 Keystone 的认证服务流程

下面以一个用户创建虚拟实例的 Keystone 认证流程来说明 Keystone 的运行机制，此流程说明了 Keystone 与其他 OpenStack 服务是如何交互和协同工作的。

首先，用户向 Keystone 提供自己的身份凭证，如用户名和密码。Keystone 会从数据库中读取数据对其进行验证，如验证通过，则向用户返回一个临时的令牌。此后用户的所有请求都会通过该令牌进行身份验证。其次，用户向 Nova 申请创建虚拟机服务，Nova 会将用户提供的令牌发给 Keystone 进行验证，Keystone 会根据令牌判断用户是否拥有进行此项操作的权限，若验证通过，则 Nova 向其提供相应的服务。最后，其他组件和 Keystone 的交互也是如此，例如，Nova 需要向 Glance 提供令牌并请求镜像，Glance 将令牌发送给 Keystone 进行验证，如果验证通过，则向 Nova 返回镜像。

值得一提的是，认证流程中还涉及服务目录和端点，具体说明如下。

（1）用户向 Keystone 提供凭证，Keystone 验证通过后向用户返回令牌，同时会返回一个通用目录（Generic Catalog）。

（2）用户使用令牌向该目录列表中的端点请求该用户对应的项目（租户）令牌，Keystone 验证通过后返回用户对应的项目（租户）列表。

（3）用户从列表中选择要访问的项目（租户），再次向 Keystone 发出请求，Keystone 验证通过后返回管理该项目（租户）的服务列表并允许访问该项目（租户）的令牌。

（4）用户会通过这个服务和通用目录映射找到服务组件的端点，通过端点找到实际服务组件的位置。

（5）用户凭借项目（租户）令牌和端点来访问实际的服务组件。

（6）服务组件会向 Keystone 提供这个用户项目令牌进行验证，Keystone 验证通过后会返回一系列的确认令牌和附加信息给服务，并执行一系列操作。

3.2.2 认证服务身份管理

OpenStack 云计算管理平台可以用来验证 Keystone 组件，其中 Dashboard 界面可用于 Keystone 身份服务的配置和管理，这里以在该平台上的操作为例进行介绍。以云管理员身份登录 Dashboard 界面，可以基于它提供的身份管理界面执行身份服务的配置和管理操作。

1. 配置 Keystone 应用环境

在安装 Keystone 服务之前需要指定用户名和密码，通过认证服务来进行身份认证。在开始阶段是没有创建任何用户的，所以必须使用授权令牌和服务的访问接口来创建特定的、用来进行身份认证的用户，之后需要创建一个管理用户的环境变量（admin-openrc.sh）来管理最终的凭证和终端。默认情况下，环境变量不会自动产生，需要用户自己创建及配置，可以在任意目录下创建，执行 Shell 命令进行操作。

当没有配置环境变量时，显示项目列表，可以看到如下操作结果。

```
[root@openstack ~]# openstack project list
Missing value auth-url required for auth plugin password
[root@openstack ~]#
```

配置环境变量，执行操作如下。

```
[root@openstack ~]# vim admin-openrc.sh
export OS_USERNAME=admin
export OS_PASSWORD='123456'
export OS_AUTH_URL=http://192.168.100.100:5000/v3
export OS_PROJECT_NAME=admin
```

```
export OS_USER_DOMAIN_NAME=Default
export OS_PROJECT_DOMAIN_NAME=Default
export OS_IDENTITY_API_VERSION=3
[root@openstack ~]#
```

显示项目列表信息，如图 3.3 所示，相关命令操作如下。

```
[root@openstack ~]# . admin-openrc.sh          //执行配置 admin-openrc.sh
[root@openstack ~]# openstack project list     //显示项目列表信息
[root@openstack ~]#
```

```
[root@openstack ~]# openstack project list
+----------------------------------+----------+
| ID                               | Name     |
+----------------------------------+----------+
| 34463c92d97c4ddcaaa69f6c7dcebdb8 | demo     |
| 77116b1383f840ac85672ce40d32acf5 | services |
| a93d65eff2d04f95bc250cf53b933c9b | admin    |
+----------------------------------+----------+
[root@openstack ~]#
```

图 3.3　显示项目列表信息

2. 控制台 Dashboard 管理

（1）项目管理

向 OpenStack 发起的任何请求必须提供项目（租户）信息。在 Dashboard 界面中，先单击左侧导航窗格中的"身份管理"主节点，再单击"项目"子节点，弹出项目界面，显示当前的项目列表信息，如图 3.4 所示。

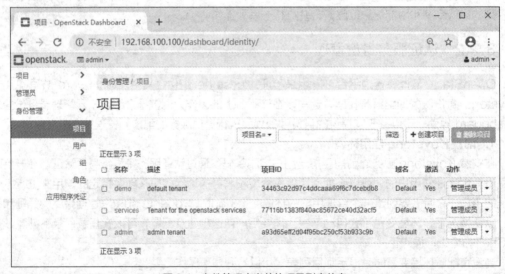

图 3.4　身份管理中当前的项目列表信息

默认情况下，项目列表中提供 3 个项目：admin、services 和 demo。项目列表中会显示每个项目的名称、描述信息、项目 ID、域名、激活状态和动作。可以在"动作"下拉列表中选择项目操作选项，默认是管理成员，还可以编辑项目、修改组、修改配额等。

（2）用户管理

OpenStack 使用了身份服务，项目通常是云计算的租户，一个租户可以是一个项目、组织或用户群。对 OpenStack 的任何请求都必须提供项目（租户）信息。云管理员能够管理整个系统的身份信息，而普通用户只能管理自己的项目（租户）的身份信息。

用户是指使用云的用户账户，包括用户名、密码和邮箱等。在 Dashboard 界面中，单击左侧导航窗格中的"身份管理"主节点，再单击"用户"子节点，会显示当前的所有用户列表信息，如图 3.5 所示。

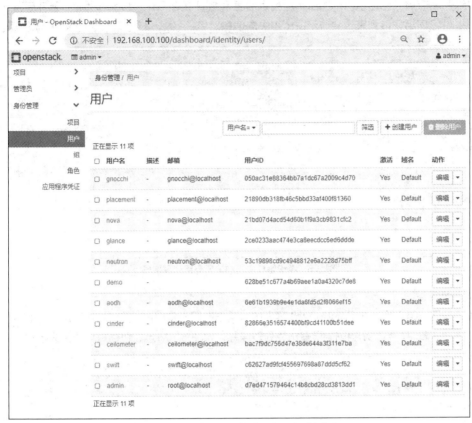

图 3.5　身份管理中当前的所有用户列表信息

可以看到，其默认提供了 11 个用户。其中，admin 和 demo 是云管理员和测试用户，其他都是 OpenStack 服务用户。该列表中会显示每个用户的用户名、描述信息、邮箱、用户 ID、激活状态、域名和动作。

一个用户至少属于一个项目，可以属于多个项目，因此至少应该添加一个项目，并为其添加用户。用户可以在这里修改密码，云管理员能够管理所有用户的密码，而普通用户在默认情况下无权修改密码，为此需要修改 Keystone 的规则文件 etc/keystone/policy.json，添加以下定义。

```
[root@openstack ~]# cat   /etc/keystone/policy.json
{}
[root@openstack ~]# vim   /etc/keystone/policy.json
{"identity:update_user":[["rule:admin_or_owner"]]}
~
"/etc/keystone/policy.json" 1L, 51C 已写入
[root@openstack ~]# cat   /etc/keystone/policy.json
{"identity:update_user":[["rule:admin_or_owner"]]}
[root@openstack ~]#
```

如果之前已经定义"identity:update_user"，则修改它即可，这样普通用户也可以修改自己的

密码。

（3）组管理

组是指用户的集合。在 Dashboard 界面中，单击左侧导航窗格中的"身份管理"主节点，再单击"组"子节点，会显示当前的所有组列表信息，如图 3.6 所示。OpenStack 云计算管理平台默认不提供任何组，该列表中会显示每个组中的信息，包括名称、描述信息、组 ID 和动作等。

图 3.6　身份管理中当前的所有组列表信息

（4）角色管理

角色表示一组权限。在 Dashboard 界面中，单击左侧导航窗格中的"身份管理"主节点，再单击"角色"子节点，会显示当前的所有角色列表信息，如图 3.7 所示，包括每个角色的名称、ID 等。

图 3.7　身份管理中当前的所有角色列表信息

OpenStack 云计算管理平台默认提供 6 个角色。其中，admin 是全局管理角色，具有最高权限；_member_ 是项目内部管理角色，表示系统的普通用户，拥有系统的正常使用和对当前项目

（租户）的管理权限，具备该角色的用户可以在项目内部创建虚拟机；ResellerAdmin 用于访问对象存储；SwiftOperator 具有访问、创建容器（Container）以及为其他用户设置访问控制列表（Access Control List，ACL）等的权限。

（5）应用程序凭证管理

在 Dashboard 界面中，单击左侧导航窗格中的"身份管理"主节点，再单击"应用程序凭证"子节点，会显示当前的所有应用程序凭证列表信息，如图 3.8 所示，包括每个应用程序凭证的名称、项目 ID、描述信息、过期信息、ID、角色和动作等。OpenStack 云计算管理平台默认没有提供任何应用程序凭证。

图 3.8　身份管理中当前的所有应用程序凭证列表信息

3.3　项目实施

3.3.1　基于 Dashboard 界面管理项目、用户、组和角色

系统管理员应当掌握命令行的使用，这在云部署过程中很重要，尤其是服务和服务用户的创建。OpenStack 管理员可以管理项目、用户和角色，可以添加、修改、删除项目和用户，将用户分配给一个或多个项目，并且修改或删除这种分配。要激活或者临时禁用一个项目和用户，可以对它们进行修改操作，也可以基于项目修改配额。在删除用户账户之前，必须从该用户的主项目中删除该用户账户。

1．项目管理配置

（1）创建项目

在 Dashboard 界面中，单击左侧导航窗格中的"身份管理"主节点，再单击"项目"子节点，弹出项目界面，单击"+创建项目"按钮，打开"创建项目"对话框，如图 3.9 所示。创建项目的名称为 new-project01，描述为 new-project01，单击"创建项目"按钮完成项目创建，如图 3.10 所示，默认情况下，新创建的项目为激活状态，即"激活"复选框已经被选中。

（2）修改项目

可以选择"动作"下拉列表中的选项来修改项目。对新建项目 new-project01 进行修改，如图 3.11 所示。

图 3.9 "创建项目"对话框

图 3.10 完成项目创建

图 3.11 修改 new-project01 项目

在"动作"下拉列表中选择"编辑项目"选项，打开"编辑项目"对话框，如图 3.12 所示，可以修改项目名称、描述，进行是否激活等相关操作。完成 new-project01 项目禁用激活显示，如图 3.13 所示。

图 3.12 "编辑项目"对话框

图 3.13 完成 new-project01 项目禁用激活显示

在"动作"下拉列表中选择"编辑项目"选项,在打开的"编辑项目"对话框中选择"项目成员"选项卡,如图 3.14 所示。可以根据需要设置项目成员,例如,单击某一用户右侧的"+"按钮可以将其加入"项目成员"列表框中,使其成为项目成员,而删除成员只需要单击"项目成员"列表框中的"-"按钮即可。还可以根据需要通过用户名右侧的"项目成员"列表框更改该项目成员的角色,一般情况下,这个值应该被设置为_member_,而管理员用户的值是 admin。admin 非常重要,是全局用户,而不是只属于某个项目,因此授予用户 admin 角色就等于赋予该用户在任何项目中管理整个云的权限。

在"动作"下拉列表中选择"查看使用量"选项,可以查看当前项目使用量情况,如图 3.15 所示。可以选择查看某一段时间的使用量,包括运行中的实例名称、vCPU 数量、磁盘、内存和时间等相关信息。

OpenStack 提供了大量配额选项,并且都是针对项目(而不是用户)的配额。在"动作"下拉列表中选择"修改配额"选项,通常包括"计算""卷"和"网络"3 个配额,配额是必须设置的,不过系统已经提供了默认值。"计算"配额配置如图 3.16 所示,"卷"配额配置如图 3.17 所示,"网络"配额配置如图 3.18 所示。

图 3.14 设置项目成员

图 3.15 查看当前项目使用量情况

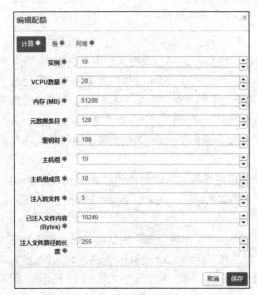

图 3.16 "计算"配额配置

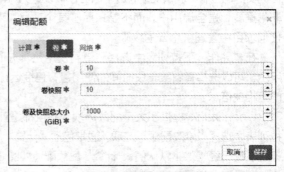

图 3.17 "卷"配额配置

图 3.18 "网络"配额配置

(3) 查看项目信息

查看 new-project01 项目，选中 new-project01 项目前面的复选框，选中项目 new-project01 名称，可以查看 new-project01 项目信息，如图 3.19 所示。其通常包括"概况""用户"和"组" 3 个选项卡。"概况"选项卡中通常包括项目名称、ID、域名、域 ID、激活和描述等相关信息。

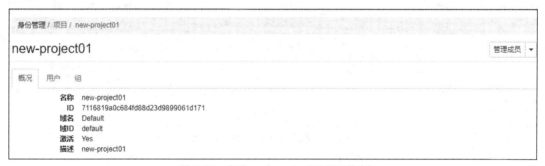

图 3.19 查看 new-project01 项目信息

查看 new-project01 项目"用户"信息，如图 3.20 所示。"用户"选项卡中通常包括用户名、描述、邮箱、用户 ID、激活、域名、角色和组角色等相关信息。

图 3.20 查看 new-project01 项目"用户"信息

查看 new-project01 项目"组"信息，如图 3.21 所示。"组"选项卡中通常包括组名称、描述、组 ID、角色等相关信息。

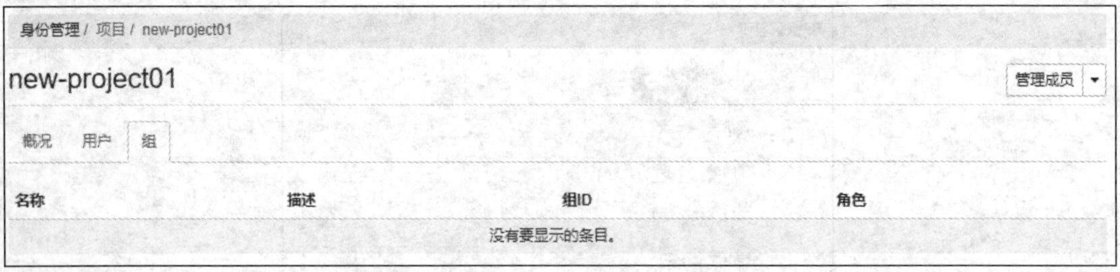

图 3.21 查看 new-project01 项目"组"信息

（4）删除项目

选中 new-project01 项目前的复选框，在"动作"下拉列表中选择"删除项目"选项，如图 3.22 所示，打开"确认删除项目"对话框，单击"删除项目"按钮即可删除项目，如图 3.23 所示。

图 3.22 删除 new-project01 项目

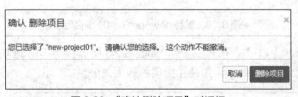

图 3.23 "确认删除项目"对话框

2. 用户管理配置

（1）创建用户

在 Dashboard 界面中，单击左侧导航窗格中的"身份管理"主节点，再单击"用户"子节点，弹出用户界面，单击"+创建用户"按钮，打开"创建用户"对话框，如图 3.24 所示。

在"创建用户"对话框中，可以设置用户信息。用户名和密码是必填信息，不过服务用户不需要密码。可以为用户指定主项目（租户），如图 3.25 所示。也可以指定用户角色，默认用户角色为_member_，如图 3.26 所示。

（2）修改用户

在"动作"下拉列表中选择"编辑"选项，对用户 user-01 进行修改，如图 3.27 所示，可以进行修改密码、禁用用户和删除用户等操作。

图 3.24 "创建用户"对话框

图 3.25 指定主项目

图 3.26 指定用户角色

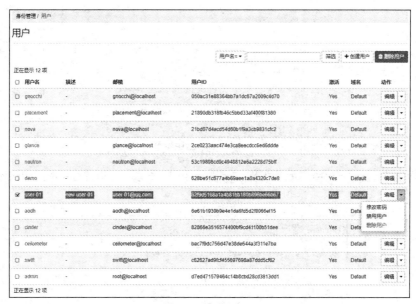

图 3.27 修改用户

在"动作"下拉列表中选择"编辑"→"修改密码"选项,可以修改用户密码,如图3.28所示。

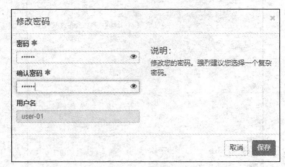

图 3.28 修改密码

在"动作"下拉列表中选择"编辑"→"禁用用户"选项,用户激活状态由"Yes"变为"No",如图3.29所示。

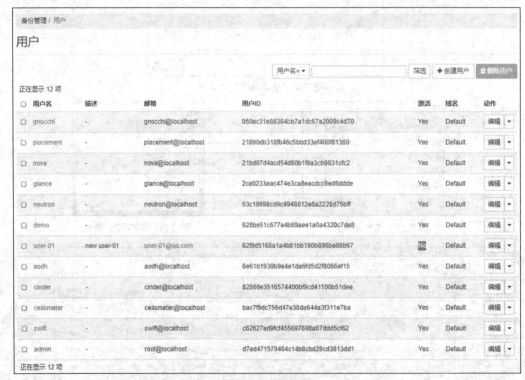

图 3.29 禁用用户

在"动作"下拉列表中选择"编辑"→"删除用户"选项,可以将选中的用户删除,如图3.30所示。

图 3.30 删除用户

(3)查看用户信息

查看 user-01 用户信息,选中 user-01 用户前的复选框,选中 user-01 用户名,可以查看 user-01 用户的"概况"信息,如图 3.31 所示。其通常包括"概况""角色分配""组"3 个选项卡。"概况"选项卡中通常包括用户名称、ID、域名、域 ID、描述、邮箱、激活、密码过期时间和主项目等相关信息。

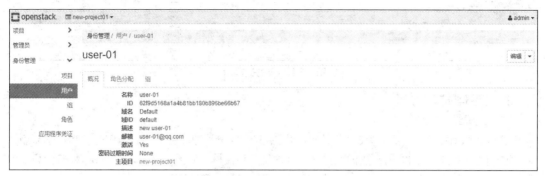

图 3.31 查看 user-01 用户的"概况"信息

查看 user-01 用户的"角色分配"信息,如图 3.32 所示。"角色分配"选项卡中通常包括项目名称、域、系统范围和角色等相关信息。查看 user-01 用户的"组"信息,如图 3.33 所示。"组"选项卡中通常包括组名称、描述和组 ID 等相关信息。

图 3.32 查看用户 user-01 的"角色分配"信息

图 3.33 查看用户 user-01 的"组"信息

查看 user-01 用户的信息,还可以在用户主界面中进行用户筛选查询,如图 3.34 所示,在主界面的"用户名="右侧的文本框中输入"user-01",单击"筛选"按钮,可以进行 user-01 用户的信息查询。

图 3.34　用户筛选查询

3. 组管理配置

（1）创建组

在 Dashboard 界面中，单击左侧导航窗格中的"身份管理"主节点，再单击"组"子节点，弹出组界面，单击"+创建组"按钮，打开"创建组"对话框，如图 3.35 所示，输入相应信息，单击"创建组"按钮，完成 workgroup-01 组的创建，如图 3.36 所示。

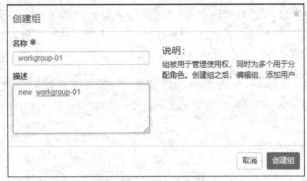

图 3.35　"创建组"对话框

图 3.36　完成 workgroup-01 组的创建

（2）更新组

在"动作"下拉列表中选择"管理成员"→"编辑组"选项，可以更新组，如图 3.37 所示。
在"动作"下拉列表中选择"管理成员"→"删除组"选项，可以删除组，如图 3.38 所示。

（3）向组中添加、删除用户

在"动作"下拉列表中选择"管理成员"选项，弹出组管理界面，如图 3.39 所示。

在组管理界面右侧单击"+添加用户"按钮，添加组成员，如图 3.40 所示，选中要添加的用户 user-01，单击"+添加用户"按钮，完成添加 user-01 组成员操作，如图 3.41 所示。

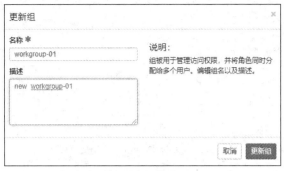

图 3.37 更新组

图 3.38 删除组

图 3.39 组管理界面

图 3.40 添加组成员

图 3.41　完成添加 user-01 组成员操作

在组管理界面中，可以单击"删除用户"按钮，打开"确认删除用户"对话框，进行用户删除操作，如图 3.42 所示。

图 3.42　删除用户

4. 角色管理配置

（1）创建角色

在 Dashboard 界面中，单击左侧导航窗格中的"身份管理"主节点，再单击"角色"子节点，弹出角色界面，单击"+创建角色"按钮，打开"创建角色"对话框，如图 3.43 所示，输入名称后单击"提交"按钮，完成 Role-01 角色的创建，如图 3.44 所示。

图 3.43　"创建角色"对话框

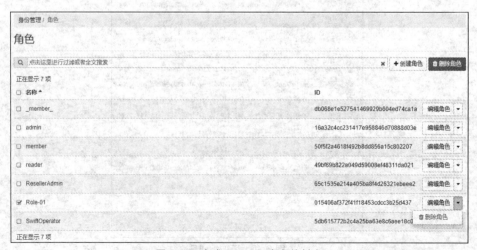

图 3.44　完成 Role-01 角色的创建

（2）编辑角色

在角色界面中，选择"编辑角色"选项，打开"编辑角色"对话框，可编辑角色名称，如图3.45所示。

图3.45 "编辑角色"对话框

（3）删除角色

在角色界面中，选择"编辑角色"→"删除角色"选项，打开"确认删除角色"对话框，如图3.46所示。单击"删除角色"按钮，即可删除选中的角色。

图3.46 "确认删除角色"对话框

3.3.2 基于命令行界面管理项目、用户、组和角色

OpenStack 管理员可以管理项目、用户、组和角色，可以添加、修改及删除项目、用户、组和角色，将用户分配给一个或多个项目，并且修改或删除这种分配。要激活或者禁用一个项目和用户，可以对它们进行修改操作，也可以基于项目修改配额。在删除用户账户之前，必须从该用户的主项目中删除该用户账户。

1. 项目管理配置

一个项目可以包括若干用户和组，每个组可以包括若干用户，每个用户可以属于不同组，对每个用户可以分配不同角色。在计算服务中，项目拥有虚拟机；在对象存储中，项目拥有容器。用户可以被关联到多个项目，每个项目和用户配对，有一个与之关联的角色。

（1）显示项目列表

通常可使用命令行来管理 OpenStack 云计算管理平台。由于命令比较多，管理员可以使用--help 命令来辅助，也可以使用管道查询命令| grep 来显示相关查询命令的使用方法。例如，执行以下命令。

V3-4 项目管理配置

```
[root@openstack ~]# openstack  project  --help
Command "project" matches:
   project create
   project delete
   project list
   project purge
```

```
        project set
        project show
[root@openstack ~]#
```

或者执行以下命令。

```
[root@openstack ~]# openstack --help | grep project
                   [--os-remote-project-name <remote_project_name> | --os-remote-
project-id <remote_project_id>]
                   [--os-remote-project-domain-name <remote_project_domain_name> |
--os-remote-project-domain-id <remote_project_domain_id>]
                   [--os-project-domain-id <auth-project-domain-id>]
                   [--os-project-name <auth-project-name>]
                   [--os-project-domain-name <auth-project-domain-name>]
                   [--os-project-id <auth-project-id>] [--os-user <auth-user>]
                        also specify the remote project option.
  --os-remote-project-name <remote_project_name>
  --os-remote-project-id <remote_project_id>
  --os-remote-project-domain-name <remote_project_domain_name>
                        Domain name of the project when authenticating to a
  --os-remote-project-domain-id <remote_project_domain_id>
                        Domain ID of the project when authenticating to a
  --os-project-domain-id <auth-project-domain-id>
                        With v3adfspassword: Domain ID containing project With
                        v3password: Domain ID containing project With v3token:
                        Domain ID containing project With v3oidcauthcode:
                        Domain ID containing project With v3samlpassword:
                        Domain ID containing project With v3totp: Domain ID
                        containing project With v3tokenlessauth: Domain ID
                        containing project With v3oidcclientcredentials:
                        Domain ID containing project With password: Domain ID
                        containing project With v3oidcaccesstoken: Domain ID
                        containing project With v3oidcpassword: Domain ID
                        containing project With token: Domain ID containing
                        project With v3applicationcredential: Domain ID
                        containing project (Env: OS_PROJECT_DOMAIN_ID)
  --os-project-name <auth-project-name>
                        project domain in v3 and ignored in v2 authentication.
                        project domain in v3 and ignored in v2 authentication.
                        and project domain in v3 and ignored in v2
                        both the user and project domain in v3 and ignored in
  --os-project-domain-name <auth-project-domain-name>
                        With v3adfspassword: Domain name containing project
                        With v3password: Domain name containing project With
                        v3token: Domain name containing project With
                        v3oidcauthcode: Domain name containing project With
                        v3samlpassword: Domain name containing project With
                        v3totp: Domain name containing project With
                        v3tokenlessauth: Domain name containing project With
                        project With password: Domain name containing project
```

```
                           With v3oidcaccesstoken: Domain name containing project
                           With v3oidcpassword: Domain name containing project
                           With token: Domain name containing project With
                           project (Env: OS_PROJECT_DOMAIN_NAME)
     --os-project-id <auth-project-id>
  endpoint add project    Associate a project to an endpoint
  endpoint group add project    Add a project to an endpoint group
  endpoint group remove project    Remove project from endpoint group
  endpoint remove project    Dissociate a project from an endpoint
  federation project list    List accessible projects
  image add project    Associate project with image
  image member list    List projects associated with image
  image remove project    Disassociate project with image
  network auto allocated topology create    Create the  auto allocated topology for project
  network auto allocated topology delete    Delete auto allocated topology for project
  network trunk create    Create a network trunk for a given project (python-neutronclient)
  project create    Create new project
  project delete    Delete project(s)
  project list    List projects
  project purge    Clean resources associated with a project
  project set    Set project properties
  project show    Display project details
  quota list    List quotas for all projects with non-default quota values or list detailed quota
informations for requested project
  quota set    Set quotas for project or class
  quota show    Show quotas for project or class. Specify "--os-compute-api-version 2.50"
or higher to see "server-groups" and "server-group-members" output for a given quota class.
  role add    Adds a role assignment to a user or group on the system, a domain, or a project
  role remove    Removes a role assignment from system/domain/project : user/group
  usage list    List resource usage per project
  usage show    Show resource usage for a single project
  vpn endpoint group list    List endpoint groups that belong to a given project
(python-neutronclient)
  vpn ike policy list    List IKE policies that belong to a given project (python-neutronclient)
  vpn ipsec policy list    List IPsec policies that belong to a given project (python-neutronclient)
  vpn ipsec site connection list    List IPsec site connections that belong to a given project
(python-neutronclient)
  vpn service list    List VPN services that belong to a given project (python-neutronclient)
[root@openstack ~]#
```

执行以下命令，可显示所有项目的 ID 和名称，包括禁用的项目，如图 3.47 所示。

```
[root@openstack ~]# openstack project list
```

（2）创建项目

执行以下命令，可创建一个名称为 new-project-02 的项目，域为 default，并查看结果，如图 3.48 所示。

```
[root@openstack ~]# openstack project create --description 'this is new-project-02' new-project-02 --domain default
[root@openstack ~]# openstack project list
```

```
[root@openstack ~]# openstack project list
+----------------------------------+--------------+
| ID                               | Name         |
+----------------------------------+--------------+
| 34463c92d97c4ddcaaa69f6c7dcebdb8 | demo         |
| 65ab3f52491247aaa908c95caaf9867a | new-project01|
| 77116b1383f840ac85672ce40d32acf5 | services     |
| a93d65eff2d04f95bc250cf53b933c9b | admin        |
+----------------------------------+--------------+
[root@openstack ~]#
```

图 3.47　显示所有项目的 ID 和名称

```
[root@openstack ~]# openstack project create --description 'this is new-project-02' new-project-02 --domain default
+-------------+----------------------------------+
| Field       | Value                            |
+-------------+----------------------------------+
| description | this is new-project-02           |
| domain_id   | default                          |
| enabled     | True                             |
| id          | edb00d51d50c428698af3c2ec1a7ca6b |
| is_domain   | False                            |
| name        | new-project-02                   |
| parent_id   | default                          |
| tags        | []                               |
+-------------+----------------------------------+
[root@openstack ~]# openstack project list
+----------------------------------+----------------+
| ID                               | Name           |
+----------------------------------+----------------+
| 34463c92d97c4ddcaaa69f6c7dcebdb8 | demo           |
| 65ab3f52491247aaa908c95caaf9867a | new-project01  |
| 77116b1383f840ac85672ce40d32acf5 | services       |
| a93d65eff2d04f95bc250cf53b933c9b | admin          |
| edb00d51d50c428698af3c2ec1a7ca6b | new-project-02 |
+----------------------------------+----------------+
[root@openstack ~]#
```

图 3.48　创建 new-project-02 项目并查看结果

（3）修改项目

修改项目需要指定项目名称或 ID，可以修改项目的名称、描述信息和激活状态等。临时禁用 new-project-02 项目，执行命令如下，并查看结果，如图 3.49 所示。

```
[root@openstack ~]# openstack project set new-project-02 --disable
[root@openstack ~]# openstack project show new-project-02
```

```
[root@openstack ~]# openstack project show new-project-02
+-------------+----------------------------------+
| Field       | Value                            |
+-------------+----------------------------------+
| description | this is new-project-02           |
| domain_id   | default                          |
| enabled     | False                            |
| id          | edb00d51d50c428698af3c2ec1a7ca6b |
| is_domain   | False                            |
| name        | new-project-02                   |
| parent_id   | default                          |
| tags        | []                               |
+-------------+----------------------------------+
[root@openstack ~]#
```

图 3.49　临时禁用 new-project-02 项目并查看结果

激活已禁用的项目 new-project-02，使用 new-project-02 的项目 ID 进行操作，执行命令如下，并查看结果，如图 3.50 所示。

```
[root@openstack ~]# openstack project set new-project-02 --enable
[root@openstack ~]# openstack project show edb00d51d50c428698af3c2ec1a7ca6b
```

```
[root@openstack ~]# openstack project set new-project-02 --enable
[root@openstack ~]# openstack project show edb00d51d50c428698af3c2ec1a7ca6b
+-------------+----------------------------------+
| Field       | Value                            |
+-------------+----------------------------------+
| description | this is new-project-02           |
| domain_id   | default                          |
| enabled     | True                             |
| id          | edb00d51d50c428698af3c2ec1a7ca6b |
| is_domain   | False                            |
| name        | new-project-02                   |
| parent_id   | default                          |
| tags        | []                               |
+-------------+----------------------------------+
[root@openstack ~]#
```

图 3.50　激活 new-project-02 项目并查看结果

修改项目的名称,将 new-project-02 项目名称修改为 project-02,项目描述修改为 this is project-02,执行以下命令,并查看结果,如图 3.51 所示。

```
[root@openstack ~]# openstack project set new-project-02 --name project-02
[root@openstack ~]# openstack project show  project-02
[root@openstack ~]# openstack project set project-02 --description 'this is project-02'
[root@openstack ~]# openstack project show  project-02
```

图 3.51　修改项目名称并查看结果

（4）删除项目

删除 project-02 项目,执行以下命令,并查看结果,如图 3.52 所示。

```
[root@openstack ~]# openstack project delete  project-02
[root@openstack ~]# openstack project list
```

图 3.52　删除 project-02 项目并查看结果

2. 用户管理配置

常用的用户管理配置命令如下。

```
[root@openstack ~]# openstack user --help
Command "user" matches:
  user create
  user delete
  user list
  user password set
  user set
  user show
[root@openstack ~]#
```

V3-5　用户管理配置

（1）显示用户列表信息

显示用户列表信息，执行以下命令，并查看用户列表信息，如图3.53所示。

```
[root@openstack ~]# openstack user list
```

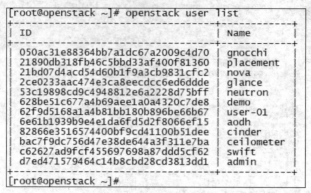

图3.53　查看用户列表信息

（2）创建用户

创建用户时必须指定用户名，还可以为用户指定项目、密码和邮件地址，建议创建用户时提供项目和密码，否则该用户不能登录Dashboard界面。执行以下命令，创建用户，并查看结果，如图3.54所示。

```
[root@openstack ~]# openstack user create --project new-project01 --password 123456 user-02
```

```
[root@openstack ~]# openstack user create --project new-project01 --password 123456 user-02
+---------------------+----------------------------------+
| Field               | Value                            |
+---------------------+----------------------------------+
| default_project_id  | 65ab3f52491247aaa908c95caaf9867a |
| domain_id           | default                          |
| enabled             | True                             |
| id                  | b4e12b98388c484e80e16686af327b4f |
| name                | user-02                          |
| options             | {}                               |
| password_expires_at | None                             |
+---------------------+----------------------------------+
[root@openstack ~]#
```

图3.54　创建用户并查看结果

（3）修改用户

修改用户需要指定用户名或ID，还可以修改用户的名称、邮件地址和激活状态等。临时禁用与激活用户账户（Dashboard界面），执行以下命令，并查看结果，如图3.55所示。

```
[root@openstack ~]# openstack user set user-02 --disable
[root@openstack ~]# openstack user show user-02
[root@openstack ~]# openstack user set user-02 --enable
[root@openstack ~]# openstack user show user-02
```

修改用户user-02名称为new-user-02，邮箱地址为new-user-02@example，执行以下命令，并查看结果，如图3.56所示。

```
[root@openstack ~]# openstack user set user-02 --name new-user-02 --email new-user-02@example
[root@openstack ~]# openstack user list
[root@openstack ~]# openstack user show new-user-02
```

```
[root@openstack ~]# openstack user set user-02 --disable
[root@openstack ~]# openstack user show user-02
+---------------------+----------------------------------+
| Field               | Value                            |
+---------------------+----------------------------------+
| default_project_id  | 65ab3f52491247aaa908c95caaf9867a |
| domain_id           | default                          |
| enabled             | False                            |
| id                  | b4e12b98388c484e80e16686af327b4f |
| name                | user-02                          |
| options             | {}                               |
| password_expires_at | None                             |
+---------------------+----------------------------------+
[root@openstack ~]# openstack user set user-02 --enable
[root@openstack ~]# openstack user show user-02
+---------------------+----------------------------------+
| Field               | Value                            |
+---------------------+----------------------------------+
| default_project_id  | 65ab3f52491247aaa908c95caaf9867a |
| domain_id           | default                          |
| enabled             | True                             |
| id                  | b4e12b98388c484e80e16686af327b4f |
| name                | user-02                          |
| options             | {}                               |
| password_expires_at | None                             |
+---------------------+----------------------------------+
[root@openstack ~]#
```

图 3.55　设置临时禁用与激活用户并查看结果

```
[root@openstack ~]# openstack user set user-02 --name new-user-02 --email new-user-02@example.com
[root@openstack ~]# openstack user list
+----------------------------------+------------+
| ID                               | Name       |
+----------------------------------+------------+
| 050ac31e88364bb7a1dc67a2009c4d70 | gnocchi    |
| 21890db318fb46c5bbd33af400f81360 | placement  |
| 21bd07d4acd54d60b1f9a3cb9831cfc2 | nova       |
| 2ce0233aac474e3ca8eecdcc6ed6ddde | glance     |
| 53c19898cd9c4948812e6a2228d75bff | neutron    |
| 628be51c677a4b69aee1a0a4320c7de8 | demo       |
| 62f9d5168a1a4b81bb180b896be66b67 | user-01    |
| 6e61b1939b9e4e1da6fd5d2f8066ef15 | aodh       |
| 82866e3516574400bf9cd41100b51dee | cinder     |
| b4e12b98388c484e80e16686af327b4f | new-user-02|
| bac7f9dc756d47e38de644a3f311e7ba | ceilometer |
| c62627ad9fcf455697698a87ddd5cf62 | swift      |
| d7ed471579464c14b8cbd28cd3813dd1 | admin      |
+----------------------------------+------------+
[root@openstack ~]# openstack user show new-user-02
+---------------------+----------------------------------+
| Field               | Value                            |
+---------------------+----------------------------------+
| default_project_id  | 65ab3f52491247aaa908c95caaf9867a |
| domain_id           | default                          |
| email               | new-user-02@example.com          |
| enabled             | True                             |
| id                  | b4e12b98388c484e80e16686af327b4f |
| name                | new-user-02                      |
| options             | {}                               |
| password_expires_at | None                             |
+---------------------+----------------------------------+
[root@openstack ~]#
```

图 3.56　修改用户名称与邮箱地址并查看结果

修改用户 user-02 的描述为 this is new-user-02，执行以下命令，并查看结果，如图 3.57 所示。

```
[root@openstack ~]# openstack user set user-02 --description 'this is new-user-02'
[root@openstack ~]# openstack user show new-user-02
```

```
[root@openstack ~]# openstack user set new-user-02 --description 'this is new-user-02'
[root@openstack ~]# openstack user show new-user-02
+---------------------+----------------------------------+
| Field               | Value                            |
+---------------------+----------------------------------+
| default_project_id  | 65ab3f52491247aaa908c95caaf9867a |
| description         | this is new-user-02              |
| domain_id           | default                          |
| email               | new-user-02@example.com          |
| enabled             | True                             |
| id                  | b4e12b98388c484e80e16686af327b4f |
| name                | new-user-02                      |
| options             | {}                               |
| password_expires_at | None                             |
+---------------------+----------------------------------+
[root@openstack ~]#
```

图 3.57　修改用户描述并查看结果

（4）删除用户

删除用户 new-user-02，执行以下命令，并查看结果，如图 3.58 所示。

图 3.58 删除用户并查看结果

3. 组管理配置

常用的组管理配置命令如下。

V3-6 组管理配置

```
[root@openstack ~]# openstack group --help
Command "group" matches:
  group add user
  group contains user
  group create
  group delete
  group list
  group remove user
  group set
  group show
```

（1）显示组列表信息

显示组列表信息，执行以下命令，并查看结果，如图 3.59 所示。

```
[root@openstack ~]# openstack group list
```

图 3.59 显示组列表信息并查看结果

（2）创建组

创建组 group-02，执行以下命令，并查看结果，如图 3.60 所示。

```
[root@openstack ~]# openstack group create --description 'this is group-02' group-02
```

图 3.60 创建组 group-02 并查看结果

（3）向组中添加用户

新建项目 project-02，创建用户 user-02，将新建用户与项目关联，执行以下命令，并查看结果，如图 3.61 所示。

```
[root@openstack ~]# openstack project create project-02 --domain default
[root@openstack ~]# openstack user create --project project-02 --password 123456 user-02
```

```
[root@openstack ~]#  openstack  project  create project-02 --domain  default
+-------------+----------------------------------+
| Field       | Value                            |
+-------------+----------------------------------+
| description |                                  |
| domain_id   | default                          |
| enabled     | True                             |
| id          | ccf4d06755bc487db185830390965f8c |
| is_domain   | False                            |
| name        | project-02                       |
| parent_id   | default                          |
| tags        | []                               |
+-------------+----------------------------------+
[root@openstack ~]# openstack user create --project project-02 --password 123456 user-02
+---------------------+----------------------------------+
| Field               | Value                            |
+---------------------+----------------------------------+
| default_project_id  | ccf4d06755bc487db185830390965f8c |
| domain_id           | default                          |
| enabled             | True                             |
| id                  | d174c2bcfa9543d1af9d76b464d7e7d1 |
| name                | user-02                          |
| options             | {}                               |
| password_expires_at | None                             |
+---------------------+----------------------------------+
```

图 3.61　新建用户与项目关联并查看结果

向组 group-02 中添加新用户 user-02，执行以下命令，并查看结果，如图 3.62 所示。

```
[root@openstack ~]# openstack group add user group-02 user-02
[root@openstack ~]# openstack group list
[root@openstack ~]# openstack group list --user user-02
```

```
[root@openstack ~]# openstack group add user group-02 user-02
[root@openstack ~]# openstack group list
+----------------------------------+--------------+
| ID                               | Name         |
+----------------------------------+--------------+
| bd56ad4b44444fa4b05f768ebeed8351 | group-02     |
| dc4187f41f104a9193bde41d874ef0df | workgroup-01 |
+----------------------------------+--------------+
[root@openstack ~]# openstack group list --user user-02
+----------------------------------+----------+
| ID                               | Name     |
+----------------------------------+----------+
| bd56ad4b44444fa4b05f768ebeed8351 | group-02 |
+----------------------------------+----------+
[root@openstack ~]#
```

图 3.62　向组 group-02 中添加新用户 user-02 并查看结果

（4）修改组信息

修改组信息，将 workgroup-01 组改名为 group-01，描述为 this is group-01，执行以下命令，并查看结果，如图 3.63 所示。

```
[root@openstack ~]# openstack group set --name group-01 --description 'this is group-01' workgroup-01
[root@openstack ~]# openstack group show group-01
```

```
[root@openstack ~]# openstack  group set --name group-01  --description 'this is group-01'  workgroup-01
[root@openstack ~]#
[root@openstack ~]#
[root@openstack ~]# openstack  group show group-01
+-------------+----------------------------------+
| Field       | Value                            |
+-------------+----------------------------------+
| description | this is group-01                 |
| domain_id   | default                          |
| id          | dc4187f41f104a9193bde41d874ef0df |
| name        | group-01                         |
+-------------+----------------------------------+
[root@openstack ~]#
```

图 3.63　修改组信息并查看结果

（5）删除组中用户及组

将用户 user-02 从组 group-02 中删除，并把组 group-02 删除，执行以下命令，并查看结果，如图 3.64 所示。

```
[root@openstack ~]# openstack group remove user group-02 user-02
[root@openstack ~]# openstack group list --user user-02
[root@openstack ~]# openstack group delete group-02
[root@openstack ~]# openstack group list
```

图 3.64 删除组中用户及组并查看结果

4. 角色管理配置

常用的角色管理配置命令如下。

```
[root@openstack ~]# openstack role --help
Command "role" matches:
  role add
  role assignment list
  role create
  role delete
  role list
  role remove
  role set
  role show
[root@openstack ~]#
```

V3-7 角色管理配置

（1）显示角色列表及查看角色信息

显示角色列表及查看角色信息，执行以下命令，并查看结果，如图 3.65 所示。

```
[root@openstack ~]# openstack role list
[root@openstack ~]# openstack role show admin
```

图 3.65 显示角色列表及查看角色信息并查看结果

（2）创建角色

用户可以是多个项目的成员，要想将用户分配给多个项目，需要定义一个角色，并将该角色分配给用户。创建一个名为 role-01 的角色，执行以下命令，并查看结果，如图 3.66 所示。

```
[root@openstack ~]# openstack role create role-01
```

```
[root@openstack ~]# openstack role create role-01
+-------------+----------------------------------+
| Field       | Value                            |
+-------------+----------------------------------+
| description | None                             |
| domain_id   | None                             |
| id          | 7663b2f1640d4134bad2dfab9ffd1334 |
| name        | role-01                          |
+-------------+----------------------------------+
[root@openstack ~]#
```

图 3.66　创建角色 role-01 并查看结果

（3）分配角色

查看当前项目、用户和角色列表相关信息，如图 3.67 所示。

```
[root@openstack ~]# openstack project list
+----------------------------------+-----------+
| ID                               | Name      |
+----------------------------------+-----------+
| 34463c92d97c4ddcaaa69f6c7dcebdb8 | demo      |
| 65ab3f52491247aaa908c95caaf9867a | project-01|
| 77116b1383f840ac85672ce40d32acf5 | services  |
| a93d65eff2d04f95bc250cf53b933c9b | admin     |
| ccf4d06755bc487db185830390965f8c | project-02|
+----------------------------------+-----------+
[root@openstack ~]# openstack user list
+----------------------------------+-----------+
| ID                               | Name      |
+----------------------------------+-----------+
| 050ac31e88364bb7a1dc67a2009c4d70 | gnocchi   |
| 21890db318fb46c5bbd33af400f81360 | placement |
| 21bd07d4acd54d60b1f9a3cb9831cfc2 | nova      |
| 2ce0233aac474e3ca8eecdcc6ed6ddde | glance    |
| 53c19898cd9c4948812e6a2228d75bff | neutron   |
| 628be51c677a4b69aee1a0a4320c7de8 | demo      |
| 62f9d5168a1a4b81bb180b896be66b67 | user-01   |
| 6e61b1939b9e4e1da6fd5d2f8066ef15 | aodh      |
| 82866e3516574400bf9cd41100b51dee | cinder    |
| bac7f9dc756d47e38de644a3f311e7ba | ceilometer|
| c62627ad9fcf455697698a87ddd5cf62 | swift     |
| d174c2bcfa9543d1af9d76b464d7e7d1 | user-02   |
| d7ed471579464c14b8cbd28cd3813dd1 | admin     |
+----------------------------------+-----------+
[root@openstack ~]# openstack role list
+----------------------------------+----------------+
| ID                               | Name           |
+----------------------------------+----------------+
| 16a32c4cc231417e958846d70888d03e | admin          |
| 49bf69b822a049d59008ef48311da021 | reader         |
| 50f5f2a4618f492b8dd856a15c802207 | member         |
| 5db615772b2c4a25ba63e8c6aee18c0a | SwiftOperator  |
| 65c1535e214a405ba8f4d26321ebeee2 | ResellerAdmin  |
| 7663b2f1640d4134bad2dfab9ffd1334 | role-01        |
| db068e1e527541469929b604ed74ca1a | _member_       |
+----------------------------------+----------------+
[root@openstack ~]#
```

图 3.67　查看当前项目、用户和角色列表相关信息

将用户指派给项目，必须将角色赋予用户与项目，这需要指定用户、角色和项目 ID。将角色分配给用户和项目，执行以下命令，并查看结果，如图 3.68 所示。

```
[root@openstack ~]# openstack role add --user user-01 --project project-01 role-01
[root@openstack ~]# openstack role assignment list --user user-01 --project project-01 --names
```

```
[root@openstack ~]# openstack role add --user user-01 --project project-01 role-01
[root@openstack ~]# openstack role assignment list --user user-01 --project project-01 --names
+---------+----------------+-------+-------------------+--------+--------+-----------+
| Role    | User           | Group | Project           | Domain | System | Inherited |
+---------+----------------+-------+-------------------+--------+--------+-----------+
| role-01 | user-01@Default |      | project-01@Default |        |        | False     |
| _member_| user-01@Default |      | project-01@Default |        |        | False     |
+---------+----------------+-------+-------------------+--------+--------+-----------+
[root@openstack ~]#
```

图 3.68　将角色分配给用户和项目并查看结果

（4）删除角色

先执行角色删除命令，再验证角色是否删除，执行以下命令，并查看结果，如图 3.69 所示。

[root@openstack ~]# openstack role remove --user user-01 --project project-01 role-01

[root@openstack ~]# openstack role list --user user-01 --project project-01

```
[root@openstack ~]# openstack role remove --user user-01 --project project-01 role-01
[root@openstack ~]# openstack role list --user user-01 --project project-01
Listing assignments using role list is deprecated. Use role assignment list --user <user-n
ame> --project <project-name> --names instead.
+----------------------------------+---------+-------------+---------+
| ID                               | Name    | Project     | User    |
+----------------------------------+---------+-------------+---------+
| db068e1e527541469929b604ed74ca1a | _member_| project-01  | user-01 |
+----------------------------------+---------+-------------+---------+
[root@openstack ~]#
```

图 3.69 删除角色并查看结果

课后习题

1. 选择题

（1）OpenStack 认证服务是通过（　　）组件来实现的。
　　A. Nova　　　　B. Keystone　　　C. Cinder　　　D. Neutron

（2）OpenStack 认证服务使用命令行方式创建角色的配置命令是（　　）。
　　A. openstack role add　　　　　B. openstack role set
　　C. openstack role create　　　　D. openstack role list

（3）OpenStack 认证服务使用命令行方式修改用户的配置命令是（　　）。
　　A. openstack user set　　　　　B. openstack role create
　　C. openstack role delete　　　　D. openstack role show

（4）OpenStack 认证服务使用命令行方式显示组的详细信息的配置命令是（　　）。
　　A. openstack group remove　　　B. openstack group set
　　C. openstack group list　　　　　D. openstack group show

2. 简答题

（1）简述 Keystone 的主要功能。

（2）简述 Keystone 的管理层次结构。

（3）简述 Keystone 的认证服务流程。

（4）如何通过 Dashboard 界面管理 OpenStack 云计算管理平台上的项目、用户、组和角色？

（5）如何通过命令行方式管理 OpenStack 云计算管理平台上的项目、用户、组和角色？

项目4
OpenStack镜像服务

【学习目标】

- 理解Glance镜像服务的基本概念。
- 理解Glance镜像服务架构以及工作流程。
- 理解镜像与实例的关系以及镜像元数据。
- 掌握管理Glance镜像的方法。

4.1 项目陈述

基于 OpenStack 构建基本的 IaaS 平台，其主要目的就是对外提供虚拟机服务。Glance 是 OpenStack 的镜像服务，它提供虚拟镜像的查询、注册和传输等服务。值得注意的是，Glance 本身并不实现对镜像的存储功能。Glance 只是一个代理，它充当镜像存储服务与 OpenStack 的其他组件（特别是 Nova）之间的纽带。在早期的 OpenStack 版本中，Glance 只有管理镜像的功能，并不具备镜像存储功能，现在 Glance 已经发展成为具有镜像上传、检索、管理和存储等多种功能的 OpenStack 核心服务。

4.2 必备知识

4.2.1 镜像服务基础

Glance 共支持两种镜像存储机制：简单文件系统机制和 Swift 服务存储镜像机制。

简单文件系统机制，是指将镜像保存在 Glance 节点的文件系统中，这种机制相对比较简单，但是存在明显的不足。例如，由于没有备份机制，当文件系统损坏时，所有的镜像都会不可用。

Swift 服务存储镜像机制，是指将镜像以对象的形式保存在 Swift 对象存储服务器中，它是 OpenStack 中用于管理对象存储的组件。Swift 具有非常可靠的备份还原机制，因此可以降低因文件系统损坏而造成的镜像不可用的风险。

镜像服务让用户能够上传和获取其他 OpenStack 服务需要使用的镜像和元数据定义等数据资产。镜像服务包括发现、注册和检索虚拟机镜像，提供了一个能够查询虚拟机镜像元数据和检索实际镜像的表述性状态传递应用程序接口（Representational State Transfer Application Programming

Interface，REST API）的软件架构风格。通过 Glance 虚拟机镜像可以存储到不同的位置，例如，从简单的文件系统到 Swift 服务这样的对象存储镜像系统。

1. 镜像与镜像服务

（1）镜像

镜像的英文为 Image，又译为映像，指一系列文件或一个磁盘驱动器的精确副本。镜像文件其实和 ZIP 压缩包类似，它将特定的一系列文件按照一定的格式制作成单一的文件，以方便用户下载和使用，如测试版的操作系统、常用工具软件等。镜像文件不仅具有 ZIP 压缩包的合成功能，它最重要的特点是可以被特

V4-1　什么是镜像

定的软件识别并可直接刻录到光盘中。其实，通常意义上的镜像文件可以再扩展一下，镜像文件中可以包含更多的信息，如系统文件、引导文件、分区表信息等，这样镜像文件就能包含一个分区甚至一块硬盘的所有信息。使用这类镜像文件的经典软件就是 Ghost，它同样具备刻录功能，不过它的刻录仅仅是将镜像文件本身保存在光盘中，而通常意义上的刻录软件可以直接将支持的镜像文件所包含的内容刻录到光盘中。Ghost 可以基于镜像文件快速安装操作系统和应用程序，还可对操作系统进行备份，当系统遇到故障不能正常启动或运行时，可快速恢复系统，使之正常工作。

用虚拟机管理程序可以模拟出一台完整的计算机，而计算机需要操作系统，此时可以将虚拟机镜像文件提供给虚拟机管理程序，让它为虚拟机安装操作系统。

虚拟磁盘为虚拟机提供了存储空间，在虚拟机中，虚拟磁盘相当于物理硬盘，即被虚拟机当作物理磁盘使用。虚拟机所使用的虚拟磁盘，实际上是一种特殊格式的镜像文件，虚拟磁盘文件用于捕获驻留在物理主机内存中的虚拟机的完整状态，并将信息以一个已明确的磁盘文件格式显示出来，每个虚拟机从其相应的虚拟磁盘文件启动，并加载到服务器内存中，随着虚拟机的运行，虚拟磁盘文件可通过更新来反映数据或状态的改变。

云环境下更加需要镜像这种高效的解决方案。镜像就是一个模板，类似 VMware 的虚拟机模板，它预先安装基本的操作系统和其他软件，例如，在 OpenStack 中创建虚拟机时，需要先准备一个镜像，再启动一个或多个该镜像实例（Instance）。整个过程已实现自动化，速度极快。如果从镜像中启动虚拟机，那么该虚拟机被删除后，镜像依然存在，但是镜像不包括本次在该虚拟机实例上的变动信息，因为镜像只是虚拟机启动的基础模板。

（2）镜像服务

镜像服务就是管理镜像，使用户能够发现、注册、获取、保存虚拟机镜像和镜像元数据。镜像数据支持多种存储系统，可以是简单文件系统，也可以是对象存储系统。在 OpenStack 中提供镜像服务的是 Glance，其主要功能如下。

V4-2　镜像服务

① 查询和获取镜像的元数据和镜像本身。

② 注册和上传虚拟机镜像，包括普通的创建、上传、下载和管理。

③ 维护镜像信息，包括元数据和镜像本身。

④ 支持通过多种方式存储镜像，包括普通的文件系统、Swift、Amazon S3 等。

⑤ 对虚拟机实例执行创建快照（Snapshot）命令来创建新的镜像，或者备份虚拟机的状态。

Glance 是关于镜像的中心，可以被终端用户或者 Nova 服务访问，接收磁盘或者镜像的 API 请求，定义镜像元数据的操作。

（3）Images API 的版本

Glance 提供的 RESTful API 目前有两个版本：Images API v1 和 Images API v2。它们存在较大差别，具体如下。

① Images API v1 只提供基本的镜像和成员操作功能，包括镜像创建、删除、下载，镜像列

表、详细信息查询和更新,以及镜像租户成员的创建、删除和列表。

② Images API v2 除了支持 Images API v1 的所有功能外,主要增加了镜像位置的添加、删除、修改,元数据和命名空间(Namespace)操作,以及镜像标记(Image Tag)操作。

这两个版本对镜像存储的支持相同,Images API v1 从 OpenStack 发行的 Newton 版本开始已经过时,迁移的路径使用 Images API v2 进行替代。按照 OpenStack 标准的弃用政策,Images API v1 最终会被废除。

(4)虚拟机镜像的磁盘格式与容器格式

在 OpenStack 中添加一个镜像到 Glance 中时,必须指定虚拟机镜像需要的磁盘格式和容器模式,虚拟设备供应商将不同的格式布局的信息保存在一个虚拟机磁盘镜像中。OpenStack 所支持的虚拟机镜像文件磁盘格式如表 4.1 所示。

表 4.1 OpenStack 所支持的虚拟机镜像文件磁盘格式

磁盘格式	说明
RAW	非结构化的磁盘镜像格式
QCOW2	QEMU 模拟器支持的可动态扩展、写时复制的磁盘格式,是 KVM 虚拟机默认使用的磁盘文件格式
VHD	通用于 VMware、Xen、VirtualBox 以及其他虚拟机管理程序
VHDX	VHD 格式的增强版本,支持更大的磁盘容量
VMDK	一种比较通用的虚拟机磁盘格式
VDI	由 VirtualBox 虚拟机监控程序和 QEMU 仿真器支持的磁盘格式
ISO	用于光盘(如 CD-ROM)数据内容的档案格式
AKI	在 Glance 中存储的 Amazon 内核格式
ARI	在 Glance 中存储的 Amazon 虚拟内存盘格式
AMI	在 Glance 中存储的 Amazon 机器格式

Glance 对镜像文件进行管理时,往往将镜像元数据装载于一个容器中,Glance 的容器格式是指虚拟机镜像采用的文件格式,该文件格式也包含关于实际虚拟机的元数据。OpenStack 所支持的镜像文件容器格式如表 4.2 所示。

表 4.2 OpenStack 所支持的镜像文件容器格式

容器格式	说明
BARE	指定没有容器和元数据封装在镜像中,如果 Glance 和 OpenStack 的其他服务没有使用容器格式的字符串,那么为了安全,建议将其设置为 BARE
OVF	开放虚拟化格式(Open Virtualization Format)
OVA	在 Glance 中存储的开放虚拟化设备格式(Open Virtualization Appliance Format)
AKI	在 Glance 中存储的 Amazon 内核格式
ARI	在 Glance 中存储的 Amazon 虚拟内存盘格式
DOCKER	在 Glance 中存储的容器文件系统的 Docker 的 TAR 档案

容器格式可以理解成虚拟机镜像添加元数据后重新打包的格式。需要注意的是,容器格式字符串目前还不能被 Glance 或其他 OpenStack 组件使用,所以如果不能确定选择哪种容器格式,那么简单地将容器格式指定为 BARE 是安全的。

(5)镜像状态

镜像状态是 Glance 管理镜像的一个重要方面,由于镜像文件都比较大,镜像从创建到成功上传至 Glance 文件系统中的过程,是通过异步任务的方式一步步完成的,Glance 为整个 OpenStack

平台提供了镜像查询服务，可以通过虚拟机镜像的状态感知某一镜像的使用情况。OpenStack 镜像状态如表 4.3 所示。

表 4.3 OpenStack 镜像状态

镜像状态	说明
queued	表示镜像已经创建和注册，这是一种初始化状态。镜像文件刚被创建时，Glance 数据库中只有其元数据，镜像数据还没有上传到数据库中
saving	表示镜像数据在上传中，这是镜像的原始数据上传到数据库的一种过渡状态
uploading	表示进行导入数据提交调用。此状态下不允许调用 PUT/file（注意，对 queued 状态的镜像执行 PUT/file 调用会将镜像置于 saving 状态，处于 saving 状态的镜像不允许 PUT/stage 调用，因此不可能对同一镜像使用两种上传方法）
importing	表示已经完成导入调用，但是镜像还未准备好使用
active	表示已经完成导入调用，成为 Glance 中的可用镜像
deactivated	表示镜像成功创建，镜像对非管理员不可用，任何非管理员用户都无权访问镜像数据。禁止下载镜像，也禁止镜像导出和镜像克隆之类的操作
killed	表示上传镜像数据出错，上传过程中发生错误，目前不可读取
deleted	表示镜像不可用，镜像将在不久后被自动删除，但是目前 Glance 仍然保留该镜像的相关信息和原始数据
pending_delete	表示镜像不可用，镜像将被自动删除。与 deleted 相似，Glance 还没有清除镜像数据，但处于该状态的镜像不可恢复

Glance 负责管理镜像的生命周期，Glance 镜像的状态转换如图 4.1 所示。在 Glance 处理镜像过程中，当从一个状态转换到下一个状态时，通常一个镜像会经历 queued、saving、active 和 deleted 等几个状态，其他状态只有在特殊情况下才会出现。注意，Images API v1 和 Images API v2 两个版本的上传失败的处理方法有所不同。

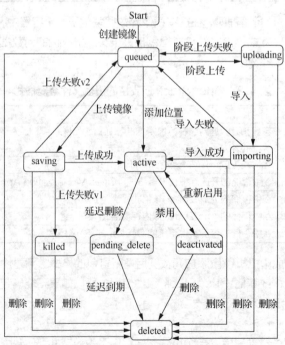

图 4.1 Glance 镜像的状态转换

（6）镜像的访问权限

① public（公有的）：可以被所有项目（租户）使用。

② private（私有的）：只能被镜像所有者所在的项目（租户）使用。

③ shared（共享的）：一个非公有的镜像可以共享给其他项目（租户），这是通过项目成员（member-*）操作来实现的。

④ protected（受保护的）：这种镜像不能被删除。

2. Glance 服务架构

Glance 镜像服务是典型的客户端/服务器（Client/Server，C/S）架构，Glance 并不负责实际的存储，只实现镜像管理功能。由于其功能比较单一，所包含的组件比较少，它主要包括 glance-api 和 glance-registry 两个子服务，如图 4.2 所示。Glance 服务器端提供一个 REST API，而使用者通过 REST API 来执行关于镜像的各种操作。

V4-3　Glance 服务架构

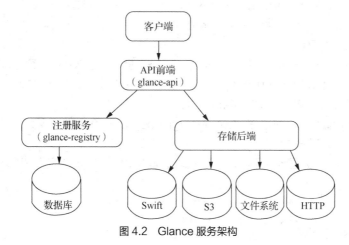

图 4.2　Glance 服务架构

（1）客户端

客户端是 Glance 服务应用程序的使用者，用于同 Glance 服务的交互和操作，可以是 OpenStack 命令行工具、Horizon 或 Nova 服务。

（2）Glance 服务进程接口（glance-api）

glance-api 是系统后台运行的服务进程，是进入 Glance 的入口，它对外提供 REST API，负责接收用户的 REST API 请求，响应镜像查询、获取和存储的调用。如果是与镜像本身存取相关的操作，则 glance-api 会将请求转发给该镜像的存储后端，通过后端的存储系统提供相应的镜像操作。

（3）Glance 注册服务进程（glance-registry）

glance-registry 是系统后台运行的 Glance 注册服务进程，负责处理与镜像元数据相关的 REST API 请求，元数据包括镜像大小、类型等信息。针对 glance-api 接收的请求，如果是与镜像的元数据相关的操作，则 glance-api 会把请求转发给 glance-registry。glance-registry 会解析请求内容，并与数据库交互，存储、处理、检索镜像的元数据。glance-api 对外提供 API，而 glance-registry 的 API 只能由 glance-api 使用。

现在的 Images API v2 已经将 glance-registry 服务集成到 glance-api 中，如果 glance-api 接收到与镜像元数据有关的请求，则会直接操作数据，无须再通过 glance-registry 服务，这样就可以减少一个中间环节。OpenStack 从 Queens 版本开始就已弃用 glance-registry 及其 API。

（4）数据库

Glance 的数据库模块存储的是镜像的元数据,可以选用 MySQL、MariaDB、SQLite 等数据库,镜像的元数据通过 glance-registry 存放在数据库中。

>
> **注意**
> 镜像本身是通过 Glance 存储驱动存放到各种存储后端的。

（5）存储后端

Glance 自身并不存储镜像,它将镜像存放在后端存储系统中,镜像本身的数据通过 glance_store（Glance 的 Store 模块,用于实现存储后端的框架）存放在各种后端中,并可从中获取。Glance 支持以下类型的存储后端。

① 本地文件存储（或者任何挂载到 glance-api 控制节点的文件系统）,这是默认配置。

② 对象存储（Object Storage）——Swift。

③ 块存储（Block Storage）——Cinder。

④ VMware 数据存储。

⑤ 分布式存储系统（Sheepdog）。既能为 QEMU 提供块存储服务,又能为支持新的存储技术即互联网 SCSI（Internet Small Computer System Interface, ISCSI）协议的客户端提供存储服务,还能支持 REST API 的对象存储服务（兼容 Swift 和 S3）。

具体使用哪种存储后端,可以在 /etc/glance/glance-api.conf 文件中配置。

3. Glance 工作流程

Glance 的工作流程如图 4.3 所示,学习这个流程可以更好地理解其工作机制。

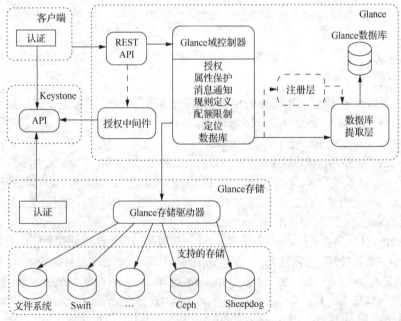

图 4.3　Glance 的工作流程

（1）流程解析

OpenStack 的操作都需要经过 Keystone 进行身份认证并授权,Glance 也不例外。Glance 是一个 C/S 架构,提供 REST API,用户通过 REST API 来执行镜像的各种操作。

Glance 域控制器是一个主要的中间件，相当于调试器，作用是将 Glance 内部服务的操作分发到以下各个功能层。

　　① 授权。其用来控制镜像的访问权限，决定镜像自己或者它的属性是否可以被修改，只有管理员和镜像的拥有者才可以执行修改操作。

　　② 属性保护（Property Protection）。这是一个可选层，只有在 Glance 的配置文件中设置时，Property Protection_file 参数才会生效。它提供两种类型的镜像属性：一种是核心属性，在镜像参数中指定；另一种是元数据属性，可以被附加到一个镜像上的任意键值对中。该层通过调用 Glance 的 public API 管理对 meta 属性的访问，也可以在配置文件中限制该访问。

　　③ 消息通知（Notifier）。将镜像变化的消息和使用镜像时发生的错误及警告添加到消息队列中。

　　④ 规则定义（Policy）。定义镜像操作的访问规则，这些规则在/etc/policy.json 文件中定义。

　　⑤ 配额限制（Quota）。如果管理员对某用户定义了镜像的上传上限，则该用户上传超过该限额的镜像时会上传失败。

　　⑥ 定位（Location）。通过 glance_store 与后台存储进行交互，例如，上传、下载镜像，管理镜像存储位置。该层能够在添加新位置时检查位置 URL 是否正确，在镜像位置改变时删除存储后端保存的镜像数据，防止镜像位置重复。

　　⑦ 数据库（DB）。实现与数据库进行交互的 API，一方面，将镜像转换为相应的格式以存储在数据库中，另一方面，将从数据库读取的信息转换为可操作的镜像对象。

　　⑧ 注册层（Registry Layer）。这是一个可选层，通过使用单独的服务控制 Glance Domain Controller 与 Glance DB 之间的安全交互。

　　⑨ Glance 数据库（Glance DB）。这是 Glance 服务使用的核心库，该库对 Glance 内部所有依赖数据库的组件来说是共享的。

　　⑩ Glance 存储（Glance Store）。其用来组织处理 Glance 和各种存储后端的交互，提供了一个统一的接口来访问后端的存储，所有的镜像文件操作都是通过调用 Glance 存储来执行的。它负责与外部存储端或本地文件存储系统的交互。

　　（2）上传镜像实例分析

　　分析完上述工作流程后，这里以上传镜像为例说明 Glance 具体的工作流程。

　　① 用户执行上传镜像命令。glance-api 服务收到请求，并通过它的中间件进行解析，获取版本号等信息。

　　② glance-registry 服务的 API 获取一个 registry client，调用 registry client 的 add_image（添加镜像）函数，此时镜像的状态为"queued"，表示该镜像 ID 已经被保留，但是镜像还未上传。

　　③ glance-registry 服务执行 registry client 的 add_image 函数，向 Glance 数据库中插入一条记录。

　　④ glance-api 调用 glance-registry 的 update_image_metadata 函数，更新数据库中该镜像的状态为"saving"，表示镜像正在上传。

　　⑤ glance-api 端存储接口提供的 add 函数上传镜像文件。

　　⑥ glance-api 调用 glance-registry 的 update_image_metadata 函数，更新数据库中该镜像的状态为"active"并发出通知，"active"表示镜像在 Glance 中完全可用。

4.2.2　镜像、实例与镜像元数据

　　虚拟机镜像包括一个持有可启动操作系统的虚拟磁盘。磁盘镜像为虚拟机文件系统提供模板，

镜像服务控制镜像存储和管理。

1. 镜像与实例的关系

实例是在云中的物理计算机节点上运行的虚拟机个体，用户可以在同一镜像中创建任意数量的实例。每个创建的实例在基础镜像（Base Image）的副本上运行，对实例的任何改变都不会影响基础镜像。快照可以抓取正在运行实例的磁盘的状态。用户可以创建快照，可以基于这些快照建立新的镜像，包括计算服务控制实例、镜像和快照的存储及管理等。

创建一个实例时必须选择一个实例类型（Flavor，也可译为类型模板或实例规格），它表示一组虚拟资源，用于定义 vCPU 的数量、可用的 RAM 和非持久化磁盘大小。用户必须从云上定义的一套可用的实例类型中进行选择。OpenStack 提供了多种预定义的实例类型，标准安装后会有 5 种默认的类型。管理员可以编辑已有的实例类型或添加新的实例类型。

可以为正在运行的实例添加或删除附加的资源，如持久性存储或公共 IP 地址。例如，OpenStack 云中一个典型的虚拟系统使用的是由 Cinder 卷服务提供的持久性存储，而不是由所选的实例类型提供的临时性存储。

未运行实例的基础镜像状态如图 4.4 所示。镜像存储拥有许多由镜像服务支持的预定义镜像，在云中，一个计算节点包括可用的 vCPU、内存和本地磁盘资源。此外，Cinder 卷服务用于存储预定义的卷。

图 4.4　未运行实例的基础镜像状态

（1）创建实例

要创建一个实例，需要选择一种镜像实例类型，以及其他可选属性，这里给出一个实例，如图 4.5 所示，所选的实例类型提供一个根卷（Root Volume，该实例中卷标为 vda）和附加的非持久性存储（该实例中卷标为 vdb）。其中 Cinder 卷服务存储映射到该实例的第 3 个虚拟磁盘（该实例中卷标为 vdc）上。

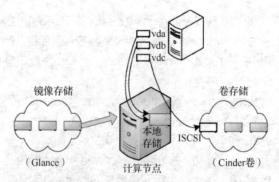

图 4.5　基于一种镜像实例类型创建的实例

镜像服务将基础镜像从镜像存储复制到本地磁盘中。本地磁盘是实例访问的第 1 个磁盘，也就是标注为 vda 的根卷。越小的实例启动速度越快，因为只有很少的数据需要通过网络复制。

创建实例时会创建一个新的非持久性空磁盘，标注为 vdb，删除该实例时，该磁盘也会被删除。

计算节点使用 ISCSI 连接到附加的 Cinder 卷存储，卷存储被映射到第 3 个磁盘（该实例中卷标为 vdc）。在计算节点上置备 vCPU 和内存资源后，该实例从根卷 vda 启动，实例运行并改变该磁盘中的数据。如果卷存储位于独立的网络，那么在存储节点配置文件中所定义的 my_block_storage_ip 选项将镜像流量指向计算节点。

具体的部署可能使用不同的后端存储或者不同的网络协议，用于卷 vda 和 vdb 的非持久性存储可能由网络存储支持而不是由本地磁盘支持。

删除实例时，除了持久性卷之外，卷存储状态也还原了，无论非持久性存储是否加密过，其都将被删除，内存和虚拟 CPU 也会被释放。在整个过程中，只有镜像本身维持不变。

（2）镜像下载工作机制

启动虚拟机之前，将虚拟机镜像服务传送到计算节点，也就是镜像下载。它的工作取决于计算节点和镜像服务的设置。

通常，计算服务会使用由调试器服务传递给它的镜像标识符（Image Identifier），并通过 Image API 请求镜像。即使镜像未存储在 Glance 中，而在一个后端（可能是对象存储、文件系统或任何其他支持的存储方式）中，也会建立从计算节点到镜像服务的连接，镜像通过该连接传输。镜像服务将镜像从后端传输到计算节点。也有可能在独立的网络中部署对象存储节点，这仍然允许镜像流量在计算节点和对象存储节点之间传输。在存储节点配置文件中配置 my_block_storage_ip 选项，允许块存储流量到达计算节点。

某些后端支持更直接的方法。收到请求后，镜像服务会返回一个直接指向后端存储的 URL，可以使用这种方式下载镜像。目前，支持直接下载的存储是文件系统存储。在计算节点的 nova.conf 配置文件的 image_file_url 中使用 filesystems 选项配置访问途径。

计算节点也可以实现镜像缓存，这意味着以前使用过的镜像不必每次都要下载。

（3）实例构建块

在 OpenStack 中，基础操作系统通常从存储在 OpenStack 镜像服务的镜像中复制，这将导致一个已知的模块状态启动的非持久性实例在关机时丢失累积的全部变化状态。也可以将操作系统放到计算系统或块存储中的持久性卷上，这将提供一个更传统的永久性系统，其累积改变的状态在重启时依然保留。要获取系统中可用的镜像，可执行以下命令。

```
openstack image list
```

2. 镜像元数据

从 OpenStack 的 Juno 发行版开始，元数据定义服务（Metadata Definition Service）就被加入 Glance。它为厂商、管理员、服务和用户提供了一个通用的 API 来自定义可用的键值对元数据，这些元数据可用于不同类型的资源，包括镜像、实例、卷、实例类型、主机聚合以及其他资源。一个定义包括一个属性的键值对、描述信息、约束和要关联的资源类型。元数据定义目录并不存储特定实例属性的值。例如，一个 vCPU 拓扑属性对核心数量的定义包括要用的基础键（如 cpu_cores）、说明信息、值约束（如要求整数值）。这样用户可以通过 Horizon 搜索这个目录，并列出能够添加到一个实例类型或镜像中的可用属性，也能在列表中看到 vCPU 拓扑属性，并知道它必须为整数。

当用户添加属性时，它的键值对会在拥有那些资源的服务中，例如，Nova 服务保存实例类型的键值对，而 Glance 保存镜像的键值对。当属性应用到不同资源类型上时，目录包括所需的其他任何附加前缀，例如，hw_用于镜像，而 hw:用于实例类型，故在一个镜像上，用户会知道将属性设置为 "hw_cpu_cores=2"。

（1）元数据的定义

元数据是描述其他数据的数据（Data about other Data），或者说是用于提供某种资源的有关信息的结构数据（Structured Data）。元数据是描述信息资源或数据等对象的数据，其使用目的在于：识别资源，评价资源，追踪资源在使用过程中的变化，实现简单高效地管理大量网络化数据，实现信息资源的有效发现、查找、一体化组织和对使用资源的有效管理。元数据的基本特点主要如下。

① 元数据一经建立，便可共享。元数据的结构和完整性依赖于信息资源的价值和使用环境，元数据的开发与利用环境往往是一个变化的分布式环境，任何一种格式都不可能完全满足不同团体的不同需要。

② 元数据是一种编码体系。元数据是用来描述数字化信息资源特别是网络信息资源的编码体系，这导致了元数据和传统数据编码体系的根本区别；元数据最为重要的特征和功能是为数字化信息资源建立了一种机器可理解框架。

元数据体系构建了电子政务的逻辑框架和基本模型，从而决定了电子政务的功能特征、运行模式和系统运行的总体性能，电子政务的运作都基于元数据来实现。其主要功能如下：描述、整合、控制和代理。

元数据也是数据，因此可以用类似数据的方法在数据库中进行存储和获取。如果提供数据元的组织同时提供描述数据元的元数据，则将会使数据元的使用变得准确而高效。用户在使用数据时可以先查看其元数据，以便获取自己所需的信息。

（2）元数据定义目录的概念体系

元数据这个术语的含义过多而且容易混淆。这里的目录是关于额外的元数据的，在多种软件和OpenStack 服务之间以自定义键值对或标记的形式传递。目前，不同 OpenStack 服务的元数据相关术语如表 4.4 所示。

表 4.4 不同 OpenStack 服务的元数据相关术语

计算节点（Nova）	块存储（Cinder）	镜像服务（Glance）
服务器（Server） • 元数据（Metadata） • 调度建议（Scheduler_hints） • 标记（Tags） 主机聚合（Host Aggregate） • 元数据（Metadata） 实例类型（Flavor） • 附加规格（Extra Specs）	快照与卷（Snapshot&Volume） • 元数据（Metadata） • 镜像元数据（Image Metadata） 卷类型（Volume Type） • 附加规格（Extra Specs） • 服务质量规则（QoS Specs）	快照与镜像（Snapshot&Image） • 属性（Properties） • 标记（Tags） • 元数据（Metadata）

元数据定义目录的概念体系如图 4.6 所示。

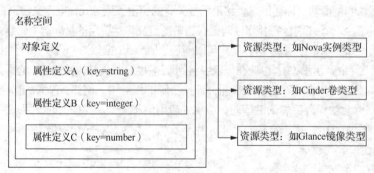

图 4.6 元数据定义目录的概念体系

一个命名空间可以关联若干种资源类型，也可以不关联任何资源类型，这对用于资源类型的 API、用户界面（User Interface，UI）是可见的。基于角色的权限访问控制（Role-Based Access Control，RBAC）权限也在命名空间中进行管理。属性可以单独定义，也可以在一个对象的上下文中定义。

（3）元数据定义目录的相关术语

① 命名空间（Name Space）。元数据定义包括命名空间中定义的任何元素指定访问控制。只允许管理员在命名空间中定义不同项目或在使用整个云时定义，可将包含的定义关联到不同的资源类型。

② 对象（Objects）。对象用于表示一组属性，包括一个或多个属性及其基本约束。组中的每个属性只能是基本类型，每个基本类型使用简单的 JavaScript 对象简谱（JavaScript Object Notation，JSON）进行定义，没有嵌套对象。对象可以定义所需要的属性，按照语义理解，一个使用该对象的用户应当提供所有需要的属性。

③ 属性（Properties）。一个属性描述了一个单一的属性及其基本约束。每个属性只能是一个基本类型，如字符串（string）、整数（integer）、数字（number）、布尔值（boolean）、数组（array）。

④ 资源类型关联（Resource Type Association）。资源类型关联定义了资源类型和适用于它们的命名空间之间的关系。这个定义可用于驱动用户界面和命令行界面（Command Line Interface，CLI）视图。例如，对象、属性和标记的同一命名空间可以用于镜像、快照、卷和实例类型，或者一个名称只能用于镜像。

值得注意的是，同一基本属性能够依据目标资源类型要求使用不同的前缀。API 根据目标资源类型，可以提供一种正确检索属性类型的方法。

（4）元数据定义示例

vCPU 拓扑可以通过元数据在镜像和实例类型上进行设置。镜像上的键与实例类型上的键有不同的前缀。实例类型上的键以 hw:为前缀，而镜像上的键以 hw_为前缀。

主机聚合即多台物理主机的集合，这个集合中的物理主机具有一个或多个硬件方面的优势，如大内存、固态磁盘等，专门用来部署数据库服务。可以制作一个镜像，在该镜像内定义好元数据，并绑定上述的主机聚合。这样，凡是用到该镜像安装系统的虚拟机，都会被指定到该集合内，并从该集合中选出一台物理机来创建虚拟机。

4.3 项目实施

4.3.1 基于 Web 界面管理镜像服务

云操作管理员将角色指定给用户，角色决定了哪些用户能够上传和管理镜像，也可以只允许管理员或操作员上传和管理镜像。可以通过 Web 界面或命令行工具来管理镜像，主要包括查看、设置和删除镜像元数据、为正在运行的实例和快照备份创建镜像。

基于部署远程桌面管理一体化 OpenStack 平台，可以验证和操作 Glance 镜像。用户以云管理员身份登录 Dashboard 界面，可以执行镜像服务管理操作。

1. 查看镜像

在 Dashboard 界面中，单击"项目"主节点，再单击"计算"→"镜像"子节点，弹出镜像列表界面，如图 4.7 所示。

图 4.7 镜像列表界面

2. 创建镜像

为了便于测试,可以从 RDO 官网下载几个专门为 OpenStack 预置的镜像文件。这里的创建镜像不是指制作镜像,而是指将已有的镜像文件上传到 Glance 中。

在镜像列表界面中单击"创建镜像"按钮,创建镜像,如图 4.8 所示,这里上传的是一个 CentOS 7.6 操作系统的镜像。可以进行镜像的描述并选择镜像源,镜像源可以选择"镜像地址"或"镜像文件"选项。选择"镜像地址"选项时,需要输入镜像所在的地址;选择"镜像文件"选项时,需要选择文件所在的位置,选择已经准备好的镜像文件,并选择合适的镜像格式,这里镜像格式为 QCOW2-QEMU 模拟器。用户根据需要设置其他选项,例如,将选项设置为"公有",表示该镜像可以被其他项目使用;将选项设置为"受保护的",表示该镜像不允许被删除;都不设置则表示镜像可以被删除。

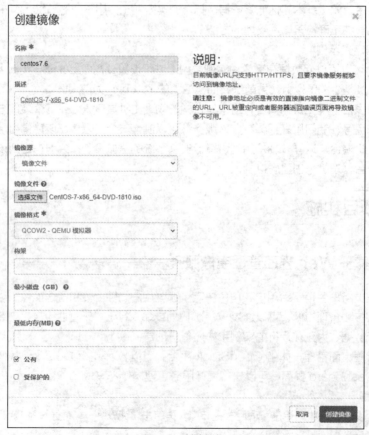

图 4.8 创建镜像

镜像通常不用特别设置元数据。单击"创建镜像"按钮，将选中的镜像文件上传到 OpenStack 中，上传成功即可完成镜像的创建。按照上述的方法，可以创建其他镜像，创建的镜像列表如图 4.9 所示。

图 4.9　创建的镜像列表

选中列表中某镜像的名称，例如，选中"centos7.6"，弹出相应界面，可查看该镜像的详细信息，如图 4.10 所示。

图 4.10　查看镜像的详细信息

3. 镜像的管理操作

镜像创建成功后，该镜像将在镜像列表中显示，可以进一步对该镜像执行管理操作，如图 4.11 所示。从界面右端的"操作"下拉列表中可以选择多种操作选项，如"创建云硬盘""编辑镜像""更新元数据""删除镜像"等。

图 4.11　执行镜像管理操作

4.3.2 基于命令行界面管理镜像服务

Glance 镜像服务可发现、注册、获取虚拟机镜像和镜像元数据，镜像数据支持多种存储系统，可以是简单文件系统、对象存储系统等。对管理员来说，使用命令行界面管理镜像的效率更高，因为使用 Web 界面上传比较大的镜像时，会长时间停留在上传的 Web 界面，所以建议使用 OpenStack 命令替代传统的 Glance 命令。在使用命令行界面之前，要加载用户的环境变量，执行如下命令。

[root@controller ~]# source /etc/keystone/admin-openrc.sh

1. 镜像管理配置命令

管理镜像可以使用 openstack 或者 glance 命令进行操作，具体可执行如下命令。

[root@controller ~]# openstack -h | grep image

或者

[root@controller ~]# glance help | grep image

使用 openstack 与 glance 命令管理镜像的格式有所不同，如图 4.12 所示。

V4-4 镜像管理配置

```
[root@controller ~]# openstack -h | grep image
                    [--os-image-api-version <image-api-version>]
  --os-image-api-version <image-api-version>
  dataprocessing image list  Lists registered images
  dataprocessing image register  Register an image
  dataprocessing image show  Display image details
  dataprocessing image tags add  Add image tags
  dataprocessing image tags remove  Remove image tags
  dataprocessing image tags set  Set image tags (Replace current image tags with provided ones)
  dataprocessing image unregister  Unregister image(s)
  image add project  Associate project with image
  image create  Create/upload an image
  image delete  Delete image(s)
  image list  List available images
  image remove project  Disassociate project with image
  image save  Save an image locally
  image set  Set image properties
  image show  Display image details
  server image create  Create a new disk image from a running server
[root@controller ~]# glance help | grep image
              [--os-image-url OS_IMAGE_URL]
              [--os-image-api-version OS_IMAGE_API_VERSION]
   image-create         Create a new image.
   image-deactivate     Deactivate specified image.
   image-delete         Delete specified image.
   image-download       Download a specific image.
   image-list           List images you can access.
   image-reactivate     Reactivate specified image.
   image-show           Describe a specific image.
   image-tag-delete     Delete the tag associated with the given image.
   image-tag-update     Update an image with the given tag.
   image-update         Update an existing image.
   image-upload         Upload data for a specific image.
   location-add         Add a location (and related metadata) to an image.
   location-delete      Remove locations (and related metadata) from an image.
   location-update      Update metadata of an image's location.
   member-create        Create member for a given image.
   member-delete        Delete image member.
   member-list          Describe sharing permissions by image.
   member-update        Update the status of a member for a given image.
 --os-image-url OS_IMAGE_URL
                        Defaults to env[OS_IMAGE_URL]. If the provided image
                        url contains a version number and `--os-image-api-
                        picked as the image api version to use.
 --os-image-api-version OS_IMAGE_API_VERSION
Run `glance --os-image-api-version 1 help` for v1 help
[root@controller ~]#
```

图 4.12 使用 openstack 与 glance 命令管理镜像的格式

可以看出使用 openstack 与 glance 命令管理镜像时，命令选项众多（可以对镜像进行创建、删除、修改、显示、上传、下载等相关操作），并且每项操作的命令选项不容易记忆，如创建镜像，可以执行如下命令查询、解释相关命令选项，以方便具体操作。

root@controller ~]# openstack help image create

或者

```
[root@controller ~]# glance help image-create
```

其命令的结果如图 4.13 和图 4.14 所示。

```
[root@controller ~]# openstack help image create
usage: openstack image create [-h]
                              [-f {html,json,json,shell,table,value,yaml,yaml}]
                              [-c COLUMN] [--max-width <integer>] [--noindent]
                              [--prefix PREFIX] [--id <id>]
                              [--container-format <container-format>]
                              [--disk-format <disk-format>]
                              [--min-disk <disk-gb>] [--min-ram <ram-mb>]
                              [--file <file>] [--volume <volume>] [--force]
                              [--protected | --unprotected]
                              [--public | --private] [--property <key=value>]
                              [--tag <tag>] [--project <project>]
                              [--project-domain <project-domain>]
                              <image-name>

Create/upload an image

positional arguments:
  <image-name>          New image name

optional arguments:
  -h, --help            show this help message and exit
  --id <id>             Image ID to reserve
  --container-format <container-format>
                        Image container format (default: bare)
  --disk-format <disk-format>
                        Image disk format (default: raw)
  --min-disk <disk-gb>  Minimum disk size needed to boot image, in gigabytes
  --min-ram <ram-mb>    Minimum RAM size needed to boot image, in megabytes
  --file <file>         Upload image from local file
  --volume <volume>     Create image from a volume
  --force               Force image creation if volume is in use (only
                        meaningful with --volume)
  --protected           Prevent image from being deleted
  --unprotected         Allow image to be deleted (default)
  --public              Image is accessible to the public
  --private             Image is inaccessible to the public (default)
  --property <key=value>
                        Set a property on this image (repeat option to set
                        multiple properties)
  --tag <tag>           Set a tag on this image (repeat option to set multiple
                        tags)
  --project <project>   Set an alternate project on this image (name or ID)
  --project-domain <project-domain>
                        Domain the project belongs to (name or ID). This can
                        be used in case collisions between project names
                        exist.

output formatters:
  output formatter options

  -f {html,json,json,shell,table,value,yaml,yaml}, --format {html,json,json,shell,table,value,yaml,yaml}
                        the output format, defaults to table
  -c COLUMN, --column COLUMN
                        specify the column(s) to include, can be repeated

table formatter:
  --max-width <integer>
                        Maximum display width, 0 to disable

json formatter:
  --noindent            whether to disable indenting the JSON

shell formatter:
  a format a UNIX shell can parse (variable="value")

  --prefix PREFIX       add a prefix to all variable names
[root@controller ~]#
```

图 4.13　openstack 创建镜像帮助命令的结果

　　openstack 命令提供了许多选项来控制镜像的创建，这里列出部分常用选项。

　　--container-format：镜像容器格式。默认格式为 BARE，可用的格式还有 AMI、ARI、AKI、docker、OVA、OVF。

　　--disk-format：镜像磁盘格式。默认格式为 RAW，可用格式还有 AMI、ARI、AKI、VHD、VMDK、QCOW2、VDI 和 ISO。

　　--min-disk：启动镜像所需的最小磁盘空间，单位是 GB。

　　--min-ram：启动镜像所需的最小内存容量，单位是 MB。

　　--file：指定上传的本地镜像文件及其路径。

　　--volume：指定创建镜像的卷。

　　--protected：表示镜像是受保护的，不能被删除。

　　--unprotected：表示镜像不受保护，可以被删除。

--public：表示镜像是公有的，可以被所有项目（租户）使用。

--private：表示镜像是私有的，只能被镜像所有者（项目或租户）使用。

--property：以键值对的形式设置属性（元数据定义），可以设置多个键值对。

--tag：设置标记，也是元数据定义的一种形式，仅用 Images API v2，也可以设置多个标记。

--project：设置镜像所属的项目，即镜像的所有者。

--project-domain：设置镜像项目所属的域。

--progress：显示命令执行进度。

```
[root@controller ~]# glance help image-create
usage: glance image-create [--architecture <ARCHITECTURE>]
                           [--protected [True|False]] [--name <NAME>]
                           [--instance-uuid <INSTANCE_UUID>]
                           [--min-disk <MIN_DISK>] [--visibility <VISIBILITY>]
                           [--kernel-id <KERNEL_ID>]
                           [--tags <TAGS> [<TAGS> ...]]
                           [--os-version <OS_VERSION>]
                           [--disk-format <DISK_FORMAT>]
                           [--os-distro <OS_DISTRO>] [--id <ID>]
                           [--owner <OWNER>] [--ramdisk-id <RAMDISK_ID>]
                           [--min-ram <MIN_RAM>]
                           [--container-format <CONTAINER_FORMAT>]
                           [--property <key=value>] [--file <FILE>]
                           [--progress]

Create a new image.

Optional arguments:
  --architecture <ARCHITECTURE>
                        Operating system architecture as specified in
                        http://docs.openstack.org/trunk/openstack-
                        compute/admin/content/adding-images.html
  --protected [True|False]
                        If true, image will not be deletable.
  --name <NAME>         Descriptive name for the image
  --instance-uuid <INSTANCE_UUID>
                        Metadata which can be used to record which instance
                        this image is associated with. (Informational only,
                        does not create an instance snapshot.)
  --min-disk <MIN_DISK>
                        Amount of disk space (in GB) required to boot image.
  --visibility <VISIBILITY>
                        Scope of image accessibility Valid values: public,
                        private
  --kernel-id <KERNEL_ID>
                        ID of image stored in Glance that should be used as
                        the kernel when booting an AMI-style image.
  --tags <TAGS> [<TAGS> ...]
                        List of strings related to the image
  --os-version <OS_VERSION>
                        Operating system version as specified by the
                        distributor
  --disk-format <DISK_FORMAT>
                        Format of the disk Valid values: ami, ari, aki, vhd,
                        vmdk, raw, qcow2, vdi, iso
  --os-distro <OS_DISTRO>
                        Common name of operating system distribution as
                        specified in http://docs.openstack.org/trunk
                        /openstack-compute/admin/content/adding-images.html
  --id <ID>             An identifier for the image
  --owner <OWNER>       Owner of the image
  --ramdisk-id <RAMDISK_ID>
                        ID of image stored in Glance that should be used as
                        the ramdisk when booting an AMI-style image.
  --min-ram <MIN_RAM>   Amount of ram (in MB) required to boot image.
  --container-format <CONTAINER_FORMAT>
                        Format of the container Valid values: ami, ari, aki,
                        bare, ovf, ova, docker
  --property <key=value>
                        Arbitrary property to associate with image. May be
                        used multiple times.
  --file <FILE>         Local file that contains disk image to be uploaded
                        during creation. Alternatively, the image data can be
                        passed to the client via stdin.
  --progress            Show upload progress bar.

Run `glance --os-image-api-version 1 help image-create` for v1 help
[root@controller ~]#
```

图 4.14　glance 创建镜像帮助命令的结果

2. 查看镜像

使用如下命令查看已有的镜像列表，如图 4.15 所示。

```
[root@controller ~]# openstack image list
```
或者
```
[root@controller ~]# glance image-list
```

```
[root@controller ~]# openstack image list
+--------------------------------------+------------+--------+
| ID                                   | Name       | Status |
+--------------------------------------+------------+--------+
| 8e6e2298-bc00-4033-8ee4-3eb126053483 | centos7.6  | active |
| 9ac41140-727c-4c52-a7bb-6880016a4ab2 | windows 10 | active |
| b422d06f-b8d0-49ee-b14c-62df4731c0d1 | cirros     | active |
+--------------------------------------+------------+--------+
[root@controller ~]# glance image-list
+--------------------------------------+------------+
| ID                                   | Name       |
+--------------------------------------+------------+
| 8e6e2298-bc00-4033-8ee4-3eb126053483 | centos7.6  |
| b422d06f-b8d0-49ee-b14c-62df4731c0d1 | cirros     |
| 9ac41140-727c-4c52-a7bb-6880016a4ab2 | windows 10 |
+--------------------------------------+------------+
[root@controller ~]#
```

图 4.15 查看已有的镜像列表

执行 openstack image show 命令进一步查看某个镜像的详细信息，可以使用镜像的 ID 或镜像的名称。例如，查看镜像 centos7.6 的详细信息，可执行如下命令。

```
[root@controller ~]# openstack image show centos7.6
```
或者
```
[root@controller ~]# openstack image show 8e6e2298-bc00-4033-8ee4-3eb126053483
```
或者
```
[root@controller ~]# glance image-show 8e6e2298-bc00-4033-8ee4-3eb126053483
```
其命令的结果如图 4.16 和图 4.17 所示。

```
[root@controller ~]# openstack image show centos7.6
+------------------+------------------------------------------------------+
| Field            | Value                                                |
+------------------+------------------------------------------------------+
| checksum         | 5b61d5b378502e9cba8ba26b6696c92a                     |
| container_format | bare                                                 |
| created_at       | 2021-01-06T14:11:05Z                                 |
| disk_format      | qcow2                                                |
| file             | /v2/images/8e6e2298-bc00-4033-8ee4-3eb126053483/file |
| id               | 8e6e2298-bc00-4033-8ee4-3eb126053483                 |
| min_disk         | 0                                                    |
| min_ram          | 0                                                    |
| name             | centos7.6                                            |
| owner            | f9ff39ba9daa4e5a8fee1fc50e2d2b34                     |
| properties       | description='CentOS-7-x86_64-DVD-1810'               |
| protected        | False                                                |
| schema           | /v2/schemas/image                                    |
| size             | 4588568576                                           |
| status           | active                                               |
| tags             |                                                      |
| updated_at       | 2021-01-06T14:12:32Z                                 |
| virtual_size     | None                                                 |
| visibility       | public                                               |
+------------------+------------------------------------------------------+
[root@controller ~]# openstack image show 8e6e2298-bc00-4033-8ee4-3eb126053483
+------------------+------------------------------------------------------+
| Field            | Value                                                |
+------------------+------------------------------------------------------+
| checksum         | 5b61d5b378502e9cba8ba26b6696c92a                     |
| container_format | bare                                                 |
| created_at       | 2021-01-06T14:11:05Z                                 |
| disk_format      | qcow2                                                |
| file             | /v2/images/8e6e2298-bc00-4033-8ee4-3eb126053483/file |
| id               | 8e6e2298-bc00-4033-8ee4-3eb126053483                 |
| min_disk         | 0                                                    |
| min_ram          | 0                                                    |
| name             | centos7.6                                            |
| owner            | f9ff39ba9daa4e5a8fee1fc50e2d2b34                     |
| properties       | description='CentOS-7-x86_64-DVD-1810'               |
| protected        | False                                                |
| schema           | /v2/schemas/image                                    |
| size             | 4588568576                                           |
| status           | active                                               |
| tags             |                                                      |
| updated_at       | 2021-01-06T14:12:32Z                                 |
| virtual_size     | None                                                 |
| visibility       | public                                               |
+------------------+------------------------------------------------------+
[root@controller ~]#
```

图 4.16 使用 openstack 命令查看镜像 centos7.6 的详细信息

```
[root@controller ~]# glance image-show 8e6e2298-bc00-4033-8ee4-3eb126053483
+------------------+--------------------------------------+
| Property         | Value                                |
+------------------+--------------------------------------+
| checksum         | 5b61d5b378502e9cba8ba26b6696c92a     |
| container_format | bare                                 |
| created_at       | 2021-01-06T14:11:05Z                 |
| description      | CentOS-7-x86_64-DVD-1810             |
| disk_format      | qcow2                                |
| id               | 8e6e2298-bc00-4033-8ee4-3eb126053483 |
| min_disk         | 0                                    |
| min_ram          | 0                                    |
| name             | centos7.6                            |
| owner            | f9ff39ba9daa4e5a8fee1fc50e2d2b34     |
| protected        | False                                |
| size             | 4588568576                           |
| status           | active                               |
| tags             | []                                   |
| updated_at       | 2021-01-06T14:12:32Z                 |
| virtual_size     | None                                 |
| visibility       | public                               |
+------------------+--------------------------------------+
[root@controller ~]#
```

图 4.17 使用 glance 命令查看镜像 centos7.6 的详细信息

3. 创建镜像

先使用 SecureFX 工具将已下载好的镜像文件 "CentOS-7-x86_64-DVD-1810.iso" 与 "cirros-0.3.3-x86_64-disk.img" 传输到 CentOS 7.6 操作系统中的/root 目录下，如图 4.18 所示，再使用 openstack 命令或 Glance 命令创建镜像。

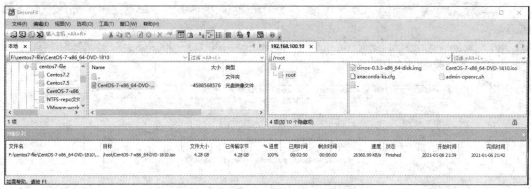

图 4.18 使用 SecureFX 工具传输镜像文件

使用 SecureFX 工具传输镜像文件完成后，可以使用命令在 CentOS 7.6 操作系统中查看镜像完成情况，如图 4.19 所示。

```
[root@controller ~]# pwd
/root
[root@controller ~]# ll
total 4494012
-rw-r--r--. 1 root root        264 Jan  6 08:30 admin-openrc.sh
-rw-------. 1 root root       1331 Dec  3  2019 anaconda-ks.cfg
-rw-r--r--. 1 root root 4588568576 Dec  9  2018 CentOS-7-x86_64-DVD-1810.iso
-rw-r--r--. 1 root root   13287936 Nov 22 22:17 cirros-0.3.3-x86_64-disk.img
[root@controller ~]#
```

图 4.19 查看镜像完成情况

创建镜像的基本命令格式如下。

openstack image create 镜像名称

使用 openstack 命令上传一个 QCOW2 格式的 "cirros-0.3.3-x86_64-disk.img" 镜像，镜像的名称为 "cirror-01"，执行如下命令。

[root@controller ~]#openstack image create --disk-format qcow2 --container-format bare < /root/cirros-0.3.3-x86_64-disk.img cirror-01

或者

[root@controller ~]#openstack image create --disk-format qcow2 --container-format bare --file /root/cirros-0.3.3-x86_64-disk.img cirror-01

其命令的结果如图 4.20 所示。

```
[root@controller ~]# openstack image create --disk-format qcow2 --container-format bare < /root/cir
ros-0.3.3-x86_64-disk.img  cirror-01
+------------------+------------------------------------------------------+
| Field            | Value                                                |
+------------------+------------------------------------------------------+
| checksum         | ee1eca47dc88f4879d8a229cc70a07c6                     |
| container_format | bare                                                 |
| created_at       | 2021-01-06T19:56:50Z                                 |
| disk_format      | qcow2                                                |
| file             | /v2/images/29ca2fc6-89d4-4771-987b-9d0fc405cc04/file |
| id               | 29ca2fc6-89d4-4771-987b-9d0fc405cc04                 |
| min_disk         | 0                                                    |
| min_ram          | 0                                                    |
| name             | cirror-01                                            |
| owner            | f9ff39ba9daa4e5a8fee1fc50e2d2b34                     |
| protected        | False                                                |
| schema           | /v2/schemas/image                                    |
| size             | 13287936                                             |
| status           | active                                               |
| tags             |                                                      |
| updated_at       | 2021-01-06T19:56:51Z                                 |
| virtual_size     | None                                                 |
| visibility       | private                                              |
+------------------+------------------------------------------------------+
[root@controller ~]#
```

图 4.20　使用 openstack 命令创建 QCOW2 格式的镜像的结果

使用 glance 命令上传一个 QCOW2 格式的"cirros-0.3.3-x86_64-disk.img"镜像，镜像的名称为"cirror-02"，执行如下命令。

[root@controller ~]#glance image-create --name "cirros-02" --disk-format qcow2 --container-format bare --progress < cirros-0.3.3-x86_64-disk.img

或者

[root@controller ~]#glance image-create --name "cirros-02" --disk-format qcow2 --container-format bare --progress --file cirros-0.3.3-x86_64-disk.img

其命令的结果如图 4.21 所示。

```
[root@controller ~]# glance image-create --name "cirros-02" --disk-format qcow2 --container-format bare --progress < cirro
s-0.3.3-x86_64-disk.img
[===============================>] 100%
+------------------+--------------------------------------+
| Property         | Value                                |
+------------------+--------------------------------------+
| checksum         | ee1eca47dc88f4879d8a229cc70a07c6     |
| container_format | bare                                 |
| created_at       | 2021-01-06T19:57:42Z                 |
| disk_format      | qcow2                                |
| id               | b0ab9ed1-6d74-4c8f-936d-365690625fe1 |
| min_disk         | 0                                    |
| min_ram          | 0                                    |
| name             | cirros-02                            |
| owner            | f9ff39ba9daa4e5a8fee1fc50e2d2b34     |
| protected        | False                                |
| size             | 13287936                             |
| status           | active                               |
| tags             | []                                   |
| updated_at       | 2021-01-06T19:57:42Z                 |
| virtual_size     | None                                 |
| visibility       | private                              |
+------------------+--------------------------------------+
[root@controller ~]#
```

图 4.21　使用 glance 命令创建 QCOW2 格式的镜像的结果

"</root/cirros-0.3.3-x86_64-disk.img"与"--file /root/cirros-0.3.3-x86_64-disk.img"命令是等价的，都用于指定所需的镜像文件所在的目录。

使用 openstack 命令上传一个 ISO 格式的"CentOS-7-x86_64-DVD-1810.iso"镜像，镜像的名称为"centos7.6-01"，执行如下命令。

[root@controller ~]# openstack image create --file /root/CentOS-7-x86_64-DVD-1810.iso --disk-format iso --container-format bare centos7.6-01

其命令的结果如图 4.22 所示。

```
[root@controller ~]# openstack image create --file /root/CentOS-7-x86_64-DVD-1810.iso --disk-format iso --container-format bare centos7.6-01
+------------------+------------------------------------------------------+
| Field            | Value                                                |
+------------------+------------------------------------------------------+
| checksum         | 5b61d5b378502e9cba8ba26b6696c92a                     |
| container_format | bare                                                 |
| created_at       | 2021-01-06T21:49:14Z                                 |
| disk_format      | iso                                                  |
| file             | /v2/images/f7b0213f-8edd-441f-b217-ecfb6754b297/file |
| id               | f7b0213f-8edd-441f-b217-ecfb6754b297                 |
| min_disk         | 0                                                    |
| min_ram          | 0                                                    |
| name             | centos7.6-01                                         |
| owner            | f9ff39ba9daa4e5a8fee1fc50e2d2b34                     |
| protected        | False                                                |
| schema           | /v2/schemas/image                                    |
| size             | 4588568576                                           |
| status           | active                                               |
| tags             |                                                      |
| updated_at       | 2021-01-06T21:51:43Z                                 |
| virtual_size     | None                                                 |
| visibility       | private                                              |
+------------------+------------------------------------------------------+
[root@controller ~]#
```

图 4.22 使用 openstack 命令创建 ISO 格式的镜像的命令

查看使用命令行界面创建的当前镜像的情况,如图 4.23 所示。

```
[root@controller ~]# openstack image list
+--------------------------------------+--------------+--------+
| ID                                   | Name         | Status |
+--------------------------------------+--------------+--------+
| f7b0213f-8edd-441f-b217-ecfb6754b297 | centos7.6-01 | active |
| b0ab9ed1-6d74-4c8f-936d-365690625fe1 | cirros-02    | active |
| 29ca2fc6-89d4-4771-987b-9d0fc405cc04 | cirror-01    | active |
| 8e6e2298-bc00-4033-8ee4-3eb126053483 | centos7.6    | active |
| 9ac41140-727c-4c52-a7bb-6880016a4ab2 | windows 10   | active |
| b422d06f-b8d0-49ee-b14c-62df4731c0d1 | cirros       | active |
+--------------------------------------+--------------+--------+
[root@controller ~]# glance image-list
+--------------------------------------+--------------+
| ID                                   | Name         |
+--------------------------------------+--------------+
| 8e6e2298-bc00-4033-8ee4-3eb126053483 | centos7.6    |
| f7b0213f-8edd-441f-b217-ecfb6754b297 | centos7.6-01 |
| 29ca2fc6-89d4-4771-987b-9d0fc405cc04 | cirror-01    |
| b422d06f-b8d0-49ee-b14c-62df4731c0d1 | cirros       |
| b0ab9ed1-6d74-4c8f-936d-365690625fe1 | cirros-02    |
| 9ac41140-727c-4c52-a7bb-6880016a4ab2 | windows 10   |
+--------------------------------------+--------------+
[root@controller ~]#
```

图 4.23 查看创建的当前镜像的情况

在 OpenStack 平台上查看当前镜像创建情况,如图 4.24 所示。

镜像

镜像名称	类型	状态	公有	受保护的	镜像格式	大小	操作
centos7.6	镜像	运行中	True	False	QCOW2	4.3 GB	创建
centos7.6-01	镜像	运行中	False	False	ISO	4.3 GB	创建
cirror-01	镜像	运行中	False	False	QCOW2	12.7 MB	创建
cirros	镜像	运行中	True	False	QCOW2	12.7 MB	创建
cirros-02	镜像	运行中	False	False	QCOW2	12.7 MB	创建
windows 10	镜像	运行中	True	False	ISO	4.1 GB	创建

正在显示 6 项

图 4.24 在 OpenStack 平台上查看当前镜像创建情况

4. 更改镜像

更改镜像的基本命令格式如下。

```
openstack image set 镜像名称
```
或者
```
glance image-update 镜像ID
```
更改镜像的选项与创建镜像的选项类似，这里介绍的更改镜像是公有的，可以被所有项目（租户）使用，默认情况下更改镜像是私有的，执行如下命令。
```
[root@controller ~]# openstack image set --public cirror-01
```
其命令的结果如图 4.25 所示，可以看到镜像"cirror-01"的 visibility 属性信息由 private 变为 public，即由私有的变为公有的。

```
[root@controller ~]# openstack image set --public cirror-01
[root@controller ~]# openstack image show cirror-01
+------------------+------------------------------------------------------+
| Field            | Value                                                |
+------------------+------------------------------------------------------+
| checksum         | ee1eca47dc88f4879d8a229cc70a07c6                     |
| container_format | bare                                                 |
| created_at       | 2021-01-06T19:56:50Z                                 |
| disk_format      | qcow2                                                |
| file             | /v2/images/29ca2fc6-89d4-4771-987b-9d0fc405cc04/file |
| id               | 29ca2fc6-89d4-4771-987b-9d0fc405cc04                 |
| min_disk         | 2                                                    |
| min_ram          | 0                                                    |
| name             | cirror-01                                            |
| owner            | f9ff39ba9daa4e5a8fee1fc50e2d2b34                     |
| protected        | False                                                |
| schema           | /v2/schemas/image                                    |
| size             | 13287936                                             |
| status           | active                                               |
| tags             |                                                      |
| updated_at       | 2021-01-09T12:00:45Z                                 |
| virtual_size     | None                                                 |
| visibility       | public                                               |
+------------------+------------------------------------------------------+
[root@controller ~]#
```

图 4.25 更改镜像属性

使用 glance image-update 命令可以更新镜像信息，当改变镜像 cirror-01 启动硬盘容量最低要求（--min-disk）时，--min-disk 默认单位为 GB，执行如下命令。
```
# glance image-update --min-disk=2 29ca2fc6-89d4-4771-987b-9d0fc405cc04
```
其命令的结果如图 4.26 所示。

图 4.26 更新镜像硬盘容量最低要求

在 OpenStack 平台上查看镜像更改情况，如图 4.27 所示。

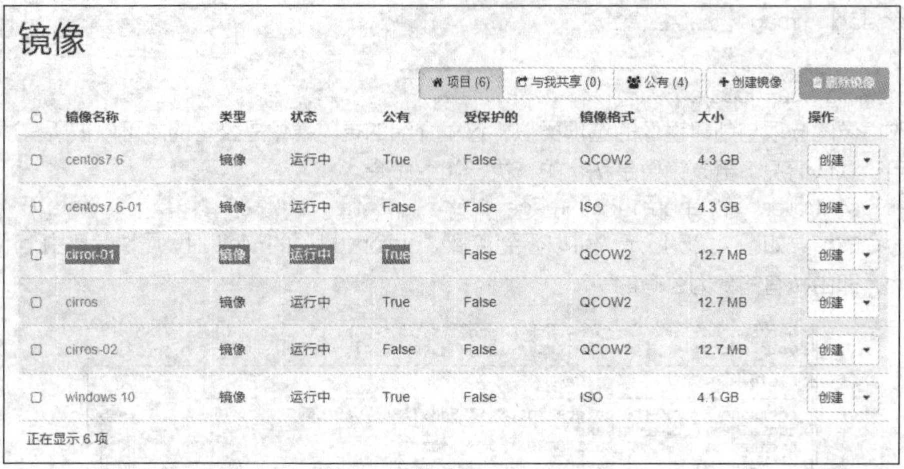

图 4.27　在 OpenStack 平台上查看镜像更改情况

5. 删除镜像

删除镜像的基本命令格式如下。

```
openstack image delete 镜像名称
```

删除镜像 "cirros-02" 可执行如下命令。

```
[root@controller ~]# openstack image delete cirros-02
[root@controller ~]# openstack image list
```

其命令的结果如图 4.28 所示。

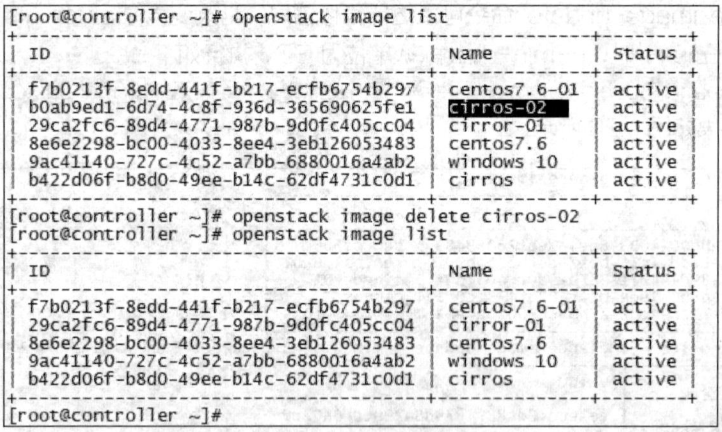

图 4.28　删除镜像

6. 镜像与项目关联

将镜像与项目关联时，需要创建镜像所属的域及项目。

创建镜像所属的域，并查看域列表信息，执行如下命令。

```
[root@controller ~]# openstack domain create --description mydomain mydomain-01
[root@controller ~]# openstack domain list
```

其命令的结果如图 4.29 所示。

创建镜像所属的项目，并与所属域相关联，执行如下命令。

```
[root@controller ~]# openstack project create --domain mydomain-01 myproject-01
```

其命令的结果如图 4.30 所示。

```
[root@controller ~]# openstack domain create --description mydomain mydomain-01
+-------------+------------------------------------+
| Field       | value                              |
+-------------+------------------------------------+
| description | mydomain                           |
| enabled     | True                               |
| id          | e7680753c4d94bb4a9cff022ea46dad1   |
| name        | mydomain-01                        |
+-------------+------------------------------------+
[root@controller ~]# openstack domain list
+----------------------------------+-------------+---------+----------------------------+
| ID                               | Name        | Enabled | Description                |
+----------------------------------+-------------+---------+----------------------------+
| 9321f21a94ef4f85993e92a228892418 | xiandian    | True    | Default Domain             |
| e2f5b3834232457997b6c6375826efd2 | heat        | True    | Stack projects and users   |
| e7680753c4d94bb4a9cff022ea46dad1 | mydomain-01 | True    | mydomain                   |
+----------------------------------+-------------+---------+----------------------------+
[root@controller ~]#
```

图 4.29 创建镜像所属的域并查看域列表信息

```
[root@controller ~]# openstack project create --domain mydomain-01 myproject-01
+-------------+----------------------------------+
| Field       | Value                            |
+-------------+----------------------------------+
| description |                                  |
| domain_id   | e7680753c4d94bb4a9cff022ea46dad1 |
| enabled     | True                             |
| id          | 78c90a8fc0de49e18f7a6b595f1bc01d |
| is_domain   | False                            |
| name        | myproject-01                     |
| parent_id   | e7680753c4d94bb4a9cff022ea46dad1 |
+-------------+----------------------------------+
[root@controller ~]#
```

图 4.30 创建镜像所属的项目并与所属域相关联

查看所创建的项目列表信息,如图 4.31 所示。

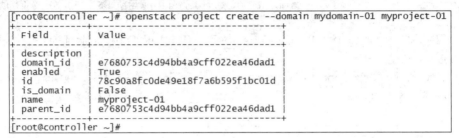

```
[root@controller ~]# openstack project list
+----------------------------------+--------------+
| ID                               | Name         |
+----------------------------------+--------------+
| 78c90a8fc0de49e18f7a6b595f1bc01d | myproject-01 |
| c88f5a1b7619420dadb4309743e53f1a | service      |
| e14b3dabf5594684913f3868669f35af | demo         |
| f9ff39ba9daa4e5a8fee1fc50e2d2b34 | admin        |
+----------------------------------+--------------+
[root@controller ~]#
```

图 4.31 查看所创建的项目列表信息

查看所创建的项目 myproject-01 的详细信息,如图 4.32 所示。

```
[root@controller ~]# openstack project show myproject-01
+-------------+----------------------------------+
| Field       | Value                            |
+-------------+----------------------------------+
| description |                                  |
| domain_id   | e7680753c4d94bb4a9cff022ea46dad1 |
| enabled     | True                             |
| id          | 78c90a8fc0de49e18f7a6b595f1bc01d |
| is_domain   | False                            |
| name        | myproject-01                     |
| parent_id   | e7680753c4d94bb4a9cff022ea46dad1 |
+-------------+----------------------------------+
[root@controller ~]#
```

图 4.32 查看所创建的项目 myproject-01 的详细信息

将镜像与项目关联的基本命令格式如下。

openstack image add project [--project-domain 项目所属域] 镜像名或 ID 项目名或 ID

将项目所属域 mydomain-01、镜像名 centos7.6-01、项目名 myproject-01 相关联,执行如下命令。

root@controller ~]# openstack image add project --project-domain mydomain-01 centos7.6-01 myproject-01

其命令的结果如图 4.33 所示。

```
[root@controller ~]# openstack image add project --project-domain mydomain-01 centos7.6-01 myproject-01
+------------+--------------------------------------+
| Field      | Value                                |
+------------+--------------------------------------+
| created_at | 2021-01-06T22:45:54Z                 |
| image_id   | f7b0213f-8edd-441f-b217-ecfb6754b297 |
| member_id  | 78c90a8fc0de49e18f7a6b595f1bc01d     |
| schema     | /v2/schemas/member                   |
| status     | pending                              |
| updated_at | 2021-01-06T22:45:54Z                 |
+------------+--------------------------------------+
[root@controller ~]#
```

图 4.33　镜像与项目关联

将镜像与项目解除关联的基本命令格式如下。

openstack image remove project [--project-domain 项目所属域] 镜像名或 ID 项目名或 ID

将项目所属域 mydomain-01、镜像名 centos7.6-01、项目名 myproject-01 解除关联，执行如下命令。

　　[root@controller ~]# openstack image remove project --project-domain mydomain-01 centos7.6-01 myproject-01

课后习题

1. 选择题

（1）创建磁盘镜像时，磁盘镜像的默认格式是（　　）。
　　A. ISO　　　　B. RAW　　　　C. QCOW2　　　　D. VDI

（2）创建磁盘镜像时，镜像容器的默认格式是（　　）。
　　A. BARE　　　B. AMI　　　　C. docker　　　　D. OVA

（3）创建磁盘镜像时，（　　）表示镜像是受保护的，不能被删除。
　　A. --public　　B. --private　　C. --protected　　D. -unprotected

（4）创建磁盘镜像时，（　　）表示镜像是公有的，可以被所有项目（租户）使用。
　　A. --public　　B. --private　　C. --protected　　D. --unprotected

（5）创建磁盘镜像时，（　　）表示显示命令执行进度。
　　A. --tag　　　　　　　　　　B. --project
　　C. --project-domain　　　　　D. --progress

（6）【多选】使用命令行查看镜像列表的命令是（　　）。
　　A. openstack image list　　　B. openstack image-list
　　C. glance image list　　　　D. glance image-list

（7）【多选】使用命令行更改镜像属性的命令是（　　）。
　　A. openstack image create 镜像名称
　　B. openstack image set 镜像名称
　　C. glance image create 镜像名称
　　D. glance image-update 镜像 ID

（8）【多选】使用命令行删除镜像的命令是（　　）。
　　A. openstack image delete 镜像名称　B. openstack image remove 镜像名称

C. glance image-delete 镜像名称　　D. glance image-remove 镜像名称

2. 简答题

（1）什么是镜像？镜像有什么作用？

（2）简述 OpenStack 镜像服务的主要功能。

（3）列举镜像文件的磁盘格式。

（4）简述 Glance 镜像状态转换过程。

（5）简述镜像的访问权限。

（6）简述 Glance 的工作流程。

（7）简述镜像和实例的关系。

（8）简述镜像元数据的作用。

项目5
OpenStack网络服务

【学习目标】

- 掌握网络虚拟化技术。
- 掌握OpenStack网络服务基础。
- 掌握Neutron主要插件、代理与服务。
- 掌握基于Web界面网络服务管理的配置方法。
- 掌握基于命令行界面网络服务管理的配置方法。

5.1 项目陈述

Neutron 是 OpenStack 最重要的网络服务资源之一，它为 OpenStack 管理所有的网络方面的虚拟网络基础架构和访问层面的物理网络基础架构。没有网络，OpenStack 将无法正常工作。在 OpenStack 中，网络、计算和存储是其核心内容，也是核心组件，可通过具体的功能实现和服务访问，提供云计算环境的虚拟网络功能。OpenStack 的网络服务最主要的功能就是为虚拟机实例提供网络连接，最初由 Nova 的一个单独模块 nova-network 实现。这种网络服务与计算服务的耦合方案并不符合 OpenStack 的特性，而且支持的网络服务有限，无法适应大规模、高密度和多项目的云计算，现在已经被专门的网络服务 Neutron 所取代。Neutron 为整个 OpenStack 环境提供软件定义网络（Software Defined Network，SDN）支持，主要功能包括二层交换、三层路由、防火墙及负载均衡等。在 OpenStack 中，网络功能是最复杂的功能，很多计算和存储方面的问题都是和网络紧密相关的。

5.2 必备知识

5.2.1 网络虚拟化

OpenStack 网络服务最核心的任务就是对二层物理网络进行抽象和管理。OpenStack 部署在 Linux 平台上，涉及 Linux 虚拟网络，它是 OpenStack 网络服务的基础。在讲解 Neutron 之前，有必要介绍一下这方面的知识。

1. Linux 网络虚拟化

传统的物理网络中，往往部署了一系列物理服务器，每台服务器上分别运行不同的服务和应用，这些服务器拥有一块或多块网卡，通过交换机连接起来，如图 5.1 所示。

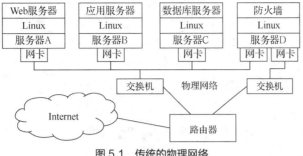

图 5.1 传统的物理网络

实现虚拟化之后，多个物理服务器可以被虚拟机取代，部署在同一台物理服务器上。虚拟机由虚拟机管理程序（Hypervisor）实现，在 Linux 操作系统中，Hypervisor 通常采用 KVM。虚拟机在对服务器进行虚拟化的同时，也对网络进行虚拟化。虚拟化的网络如图 5.2 所示。

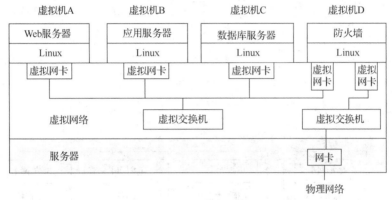

图 5.2 虚拟化的网络

Hypervisor 为虚拟机创建了一块或多块虚拟网卡，虚拟网卡等同于虚拟机的物理网卡。物理交换机在虚拟网络中被虚拟为虚拟交换机，虚拟机的虚拟网卡连接在虚拟交换机上，虚拟交换机再通过物理主机的物理网卡连接到外部网络。对于简单的物理网络来说，其主要工作是网络接口（网卡）和交换设备（交换机）的虚拟化。

2. Linux 虚拟网桥

与物理机不同，虚拟机并没有硬件设备与物理机和其他虚拟机进行通信。Linux KVM 的解决方案是提供虚拟网桥设备，像物理交换机具有若干网络接口一样，在网桥上创建多个虚拟的网络接口，每个网络接口再与 KVM 虚拟机的网卡相连。

Linux 虚拟网桥示意图如图 5.3 所示，其进一步说明了虚拟网卡、虚拟网桥、物理网卡与物理交换机的关系。在 Linux 的 KVM 虚拟系统中，为支持虚拟机的网络通信，网桥接口的名称通常以 vnet 开头，加上从 0 开始的数字编号，如 vnet0、vnet1，在创建虚拟机时会自动创建这些接口。虚拟网桥 br1 和 br2 分别连接到物理主机的物理网卡 1 和网卡 2。br1 上的网桥接口 vnet0 连接到虚拟机 A 的虚拟网卡，但是它所连接的物理网卡未连接到物理交换机，因此虚拟机 A 不能与外部网络通信。br2 上的两个虚拟网桥接口 vnet1 和 vnet2 分别连接到虚拟机 B 和虚拟机 C 的虚拟网卡上，而 br2 所连接的物理网卡 2 又连接到外部的物理交换机上，因此虚拟机 B 和虚拟机 C 可以连接到 Internet。br3 上的虚拟网桥接口 vnet3 连接到虚拟机 D 的虚拟网卡，但是它不与任何物理网卡连接，无法访问物理主机和外部网络。

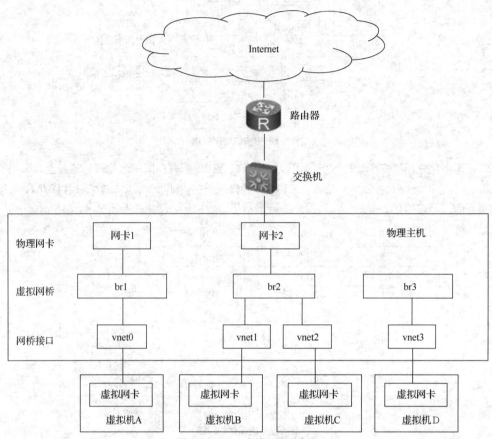

图 5.3　Linux 虚拟网桥示意图

3. 虚拟局域网

一个网桥可以连接若干虚拟机。当多个虚拟机连接在同一网桥上时，每个虚拟机发出的广播报文会引发广播风暴，影响虚拟机的网络性能。通常使用虚拟局域网（Virtual Local Area Network，VLAN）将部分虚拟机的广播报文限制在特定范围内，不影响其他虚拟机的网络通信。将多个虚拟机划分到不同的 VLAN 中，同一 VLAN 的虚拟机相当于连接在同一网桥上，而不同 VLAN 之间的通信会被隔离。目前，交换机广泛应用 VLAN 技术来隔离不同网络，以提高网络的安全和性能。在 Linux 虚拟化环境中，通常会将网桥与 VLAN 对应起来，也就是将网桥划分到不同的 VLAN 中。

VLAN 将一个物理交换机在逻辑上分成多个交换机，限制了广播的范围，在网络第二层将计算机隔离到不同的 VLAN 中。VLAN 使用 802.1Q 协议，该协议将每个虚拟机发出的包加上一个标签，一个局域网可以有 4096 个标签，每个标签可被看作一个独立的 VLAN，每个虚拟机只能接收与自身设置相同 VLAN 号码的数据报文。网桥与 VLAN 如图 5.4 所示。

VLAN 是具有 802.1Q 标签的网络，虚拟机 C1 与虚拟机 B2 虽然位于不同的物理主机，但是这两个虚拟机的每个二层数据报文都有一个 VLAN2 标签，彼此可以通信，它们发出的二层广播报文被限制在拥有 VLAN2 标签的虚拟机之间，不会发送到其他计算机上。VLAN 的隔离是二层网络上的隔离，在三层网络层上是可以通过路由器实现三层互通的。

通常 VLAN 交换机的端口有两种配置模式——接入（Access）和骨干（Trunk），相应的端口分别称为接入端口和骨干端口。

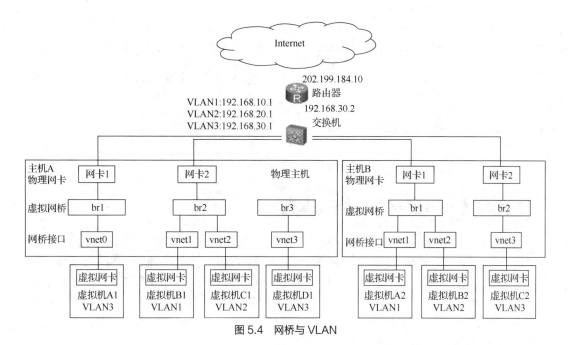

图 5.4　网桥与 VLAN

接入端口就是将数据报文"打上"VLAN 标签的端口，标签表明该端口属于哪个 VLAN，不同 VLAN 用 VLAN ID 来区分，VLAN ID 为 0~4095。接入端口都是直接与计算机网卡相连的，这样从该网卡出来的数据报文流经该端口后就被"打上"了所在 VLAN 的标签。一个接入端口只能属于一个 VLAN。骨干端口就是不处理 VLAN 标签，只转发数据报文的端口，可以允许所定义的所有 VLAN 数据报文通过。另外，不同 VLAN 之间的通信可以通过路由器来实现，如果要限制 VLAN 之间的通信，则可以在交换机中配置访问控制列表。

支持 VLAN 的物理交换机可在同一 VLAN 端口转发数据报文，在不同 VLAN 端口之间隔离数据报文。在 Linux 虚拟化环境中，将 Linux 网桥和 VLAN 结合起来，可实现具有类似功能的虚拟交换机；将同一 VLAN 的设备挂载到同一个网桥上，这些设备之间即可通信；可使用 VLAN 设备隔离不同 VLAN 之间的通信。这些在 OpenStack 中可以轻松实现。

4．开放虚拟交换机

开放虚拟交换机（Open vSwitch，OVS）与硬件交换机具备相同特性，可在不同虚拟化平台之间移植，适合在生产环境中部署。交换设备的虚拟化对虚拟网络来说至关重要，Linux 网桥和 VLAN 的结合已经可以胜任虚拟交换机角色，Open vSwitch 则具有更多的优势。在传统的数据中心中，管理员可以对物理交换机进行配置，控制服务器的网络接入，实现网络隔离、流量监控、QoS 配置、流量优化等目标。而在云环境中，单靠物理交换机的支持，是无法区分所接的物理网卡上流经的数据报文究竟属于哪个虚拟机、哪个操作系统、哪个用户的。采用 Open vSwitch 技术的虚拟交换机可使虚拟网络的管理、网络状态和流量的监控得以轻松实现。

Linux 网桥和 VLAN 的结合只能实现基本的二层交换和隔离，而 Open vSwitch 不仅支持这些基本的二层交接机功能，还能将接入 Open vSwitch 的各个虚拟机分配到不同的 VLAN 中进行隔离。

Open vSwitch 支持 Open Flow，可以接受 Open Flow 控制器的管理，它可以更好地与软件定义网络融合。

Open vSwitch 能在云环境中的虚拟化平台上实现分布式虚拟交换机，如图 5.5 所示，可以将不同物理主机上的 Open vSwitch 连接起来，形成一个大规模的虚拟网络。

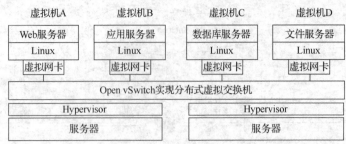

图 5.5 分布式虚拟交换机

5.2.2 OpenStack 网络服务基础

OpenStack 网络服务提供了一个 API 让用户在云中建立和定义网络连接。该网络服务的项目名称是 Neutron。OpenStack 网络负责创建和管理虚拟网络基础架构，包括网络、交换机、路由器和子网等，它们可由 OpenStack 计算服务 Nova 管理。它还提供类似防火墙的高级服务。OpenStack 网络整体上是独立的，能够部署到专用主机上。如果部署中使用控制节点主机来运行集中式计算组件，则可以将网络服务部署到特定主机上。OpenStack 网络组件可与身份服务、计算服务、仪表板等多个 OpenStack 组件进行整合。

1. Neutron 网络结构

Neutron 网络的目的是灵活地划分物理网络。OpenStack 所在的整个物理网络都会由 Neutron"池化"为网络资源池，Neutron 对这些网络资源进行处理，为项目（租户）提供独立的虚拟网络环境。Neutron 创建各种资源对象并进行连接和资源整合，从而形成项目（租户）的私有网络。一个简化的 Neutron 网络架构如图 5.6 所示，其包括一个外部网络（External Network）、一个内部网络（Internal Network）和一个路由器（Router）。

V5-1 Neutron 网络结构

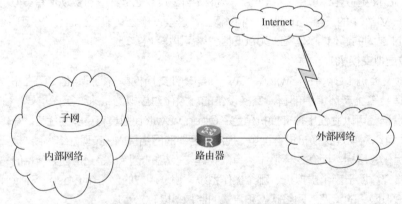

图 5.6 一个简化的 Neutron 网络架构

外部网络负责连接 OpenStack 项目之外的网络环境，如 Internet。与其他网络不同，外部网络不仅仅是一个虚拟网络，更重要的是，它表示 OpenStack 网络能被外部物理网络接入并访问。外部网络可能是企业的局域网，也可能是互联网，这类网络并不是由 Neutron 直接管理的。

内部网络完全由软件定义，又称私有网络（Private Network）。它是虚拟机实例所在的网络，能够直接连接到虚拟机。项目（租户）用户可以创建自己的内部网络。默认情况下，项目（租户）

之间的内部网络是相互隔离的,不能共享。内部网络由 Neutron 直接配置和管理。

路由器用于将内部网络与外部网络连接起来,因此,要使虚拟机访问外部网络,必须创建一个路由器。

Neutron 需要实现的主要是内部网络和路由器。内部网络是对二层网络的抽象,模拟物理网络的二层局域网,对于项目来说,它是私有的。路由器则是对三层网络的抽象,模拟物理路由器,为用户提供路由、网络地址转换(Network Address Translation,NAT)等服务。

2. Neutron 管理的网络资源

Neutron 网络中的子网并非模拟物理网络的子网,而属于三层网络的组成部分,用于描述 IP 地址范围。Neutron 使用网络、子网和端口等术语来描述所管理的网络资源。

V5-2 Neutron 管理的网络资源

(1)网络。网络指一个隔离的二层广播域,类似于交换机中的 VLAN。Neutron 支持多种类型的网络,如 Flat、VLAN、VXLAN 等。

(2)子网。子网指一个 IPv4 或者 IPv6 的地址段及其相关配置状态。虚拟机实例的 IP 地址从子网中分配,每个子网需要定义 IP 地址的范围和掩码。

(3)端口。端口指连接设备的连接点,类似于虚拟交换机上的一个网络端口。端口定义了 MAC 地址和 IP 地址,当虚拟机的虚拟网卡绑定到端口时,端口会将 MAC 地址和 IP 地址分配给该虚拟网卡。

通常可以创建和配置网络、子网和端口来为项目(租户)搭建虚拟网络。网络必须属于某个项目(租户),一个项目中可以创建多个网络,但是不能重复。一个端口必须属于某个子网,一个子网可以有多个端口,一个端口可以连接一个虚拟机的虚拟网卡。不同项目的网络设置可以重复,可以使用同一类型或范围的 IP 地址。

3. Neutron 网络拓扑类型

用户可以在自己的项目内创建用于连接的项目网络。默认情况下,这些项目网络是彼此隔离的,不能在项目之间共享。OpenStack 网络服务 Neutron 支持以下类型的网络隔离和叠加(Overlay)技术,即网络拓扑类型。

V5-3 Neutron 网络拓扑类型

(1)Local。Local 网络与其他网络和节点隔离。该网络中的虚拟机实例只能与位于同一节点上同一网络的虚拟机实例通信,实际意义不大,主要用于测试环境。位于同一 Local 网络的实例可以通信,位于不同 Local 网络的实例无法通信。一个 Local 网络只能位于一个物理节点,无法跨节点部署。

(2)Flat。Flat 是一种简单的扁平网络拓扑,所有虚拟机实例都连接在同一网络中,能与位于同一网络的实例进行通信,并且可以跨多个节点。这种网络不使用 VLAN,没有对数据报文"打上" VLAN 标签,无法进行网络隔离。Flat 是基于不使用 VLAN 的物理网络实现的虚拟网络,每个物理网络最多只能实现一个虚拟网络。

(3)VLAN。VLAN 是支持 802.1Q 协议的网络,使用 VLAN 标签标记数据报文,实现网络隔离。同一 VLAN 网络中的实例可以通信,不同 VLAN 网络中的实例只能通过路由器来通信。VLAN 网络可以跨节点,是应用最广泛的网络拓扑类型之一。

(4)VXLAN。VXLAN(Virtual Extensible LAN)可以看作 VLAN 的一种扩展,相比于 VLAN,它有更大的扩展性和灵活性,是目前支持大规模多项目网络环境的解决方案。由于 VLAN 的封包头部限制长度为 12 位,导致 VLAN 的数量限制是 4096(2^{12})个,无法满足网络空间日益增长的需求。目前,VXLAN 的封包头部有 24 位用作 VXLAN 标识符(VNID)来区分 VXLAN 网段,最多可以支持 16777216(2^{24})个网段。

VLAN 使用生成树协议(Spanning Tree Protocol,STP)防止环路,导致一半的网络路径被阻断。VXLAN 的数据报文是封装到 UDP 通过三层传输和转发的，可以完整地利用三层路由，能克服 VLAN 和物理网络基础设施的限制，更好地利用已有的网络资源。

（5）GRE。通用路由封装（Generic Routing Encapsulation，GRE）是用一种网络层协议去封装另一种网络层协议的隧道技术。GRE 的隧道由两端的源 IP 地址和目的 IP 地址定义，它允许用户使用 IP 封装 IP 等协议，并支持全部的路由协议。在 OpenStack 环境中使用 GRE 意味着"IP over IP"，即对 IP 数据流使用 GRE 隧道技术，GRE 与 VXLAN 的主要区别在于，它是使用 IP 而非 UDP 进行封装的。

（6）GENEVE。通用网络虚拟封装（Generic Network Virtualization Encapsulation，GENEVE）的目标宣称是仅定义封装数据格式，尽可能实现数据格式的弹性和扩展性。GENEVE 封装的包通过标准的网络设备传送，即通过单播或多播寻址，包从一个隧道端点传送到另一个或多个隧道端点。GENEVE 帧格式由一个封装在 IPv4 或 IPv6 的 UDP 中的简化的隧道头部组成。GENEVE 的推出主要是为了解决封装时添加的元数据信息问题，以适应各种虚拟化场景。

> **注意** 随着云计算、物联网、大数据、人工智能等技术的普及，网络虚拟化技术的趋势为在传统单层网络基础上叠加一层逻辑网络。这将网络分成两个层次，传统单层网络称为承载网络（Underlay 网络），叠加在其上的逻辑网络称为叠加网络或覆盖网络（Overlay 网络）。Overlay 网络的节点通过虚拟网络或逻辑网络的连接进行通信，每一个虚拟的或逻辑的连接对应于 Underlay 网络的一条路径（Path），由多条前后衔接的连接组成。Overlay 网络无须对基础网络进行大规模修改，不用关心这些底层实现，是实现云网络整合的关键，VXLAN、GRE 和 GENEVE 都是基于隧道技术的 Overlay 网络。

4. Neutron 基本架构

与 OpenStack 的其他服务和组件的设计思想一样，Neutron 也采用分布式架构，由多个组件（子服务）共同对外提供网络服务，其基本架构如图 5.7 所示。

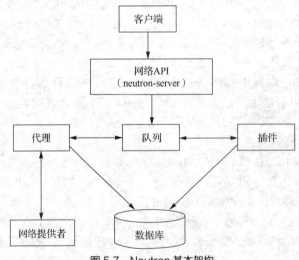

图 5.7 Neutron 基本架构

Neutron 基本架构非常灵活，层次较多，既支持各种现在或者将来会出现的先进网络技术，又支持分布式部署，以获得足够的扩展性。

Neutron 仅有一个主要服务进程，即 neutron-server。它运行于控制节点上，对外提供 OpenStack 网络 API 作为访问 Neutron 的入口，收到请求后调用插件进行处理，最终由计算节点和网络节点上的各种代理完成请求。

网络提供者是指提供 OpenStack 网络服务的虚拟或物理网络设备，如 Linux Bridge、Open vSwitch，或者其他支持 Neutron 的物理交换机。

与其他服务一样，Neutron 的各组件服务之间需要相互协调和通信，网络 API（neutron-server）、插件和代理之间通过消息队列进行通信（默认使用 RabbitMQ 实现）和相互调用。

数据库（默认使用 MariaDB）用于存放 OpenStack 的网络状态信息，包括网络、子网、端口、路由器等。

客户端是指使用 Neutron 服务的应用程序，可以是命令行工具（脚本）、Horizon（OpenStack 图形操作界面）和 Nova 计算服务等。

neutron-server 提供一组 API 来定义网络连接和 IP 地址，供 Nova 等客户端调用，它本身也是基于层次模型设计的。

Neutron 遵循 OpenStack 的设计原则，采用开放性架构，通过插件、代理与网络提供者的配合来实现各种网络功能。插件是 Neutron 的一种 API 的后端实现，目的是增强扩展性。插件按照功能可以分为 Core Plugin 和 Service Plugin 两种类型。Core Plugin 提供基础二层虚拟网络技术，实现网络、子网和端口等核心资源的抽象。Service Plugin 是指 Core Plugin 之外的其他插件，提供路由器、防火墙、安全组、负载均衡等服务。值得一提的是，直到 OpenStack 的 Havana 版本，Neutron 才开始提供一个名为 L3 Router Service Plugin 的插件支持路由器。

插件由 neutron-server 的 Core Plugin API 和 Extension Plugin API 调用，用于确定具体的网络功能，即要配置什么样的网络。插件处理 neutron-server 发送来的请求，主要职责是在数据库中维护 Neutron 网络的状态信息，通知相应的代理实现具体的网络功能。每一个插件支持一组 API 资源并完成特定的操作，这些操作最终由远程过程调用（Remote Procedure Call，RPC）插件调用相应的代理来完成。

代理处理插件转来的请求，负责在网络中实现各种网络功能。代理使用物理网络设备或虚拟化技术完成实际的操作任务，如用于路由器具体操作的 L3 Agent。

插件、代理与网络提供者配套使用，例如，网络提供者是 Linux Bridge，就需要使用 Linux Bridge 的插件和代理；如果换成 Open vSwitch，则需要改成相应的插件和代理。

5. Neutron 的物理部署

Neutron 与其他 OpenStack 服务组件协同工作，可以部署在多个物理主机节点上，主要涉及控制节点、网络节点和计算节点，每类节点可以部署多个。

（1）控制节点和计算节点

控制节点上可以部署 neutron-server（API）、Core Plugin 和 Service Plugin 代理。这些代理包括 neutron-plugin-agent、neutron-metadata、neutron-dhcp-agent、neutron-l3-agent、neutron-lbaas-agent 等。Core Plugin 和 Service Plugin 已经集成到 neutron-server 中，不需要运行独立的 Plugin 服务。

控制节点和计算节点需要部署 Core Plugin 的代理，因为控制节点与计算节点只有通过该代理才能建立二层连接。

（2）控制节点和网络节点

可以通过增加网络节点承担更大的负载。该方案特别适用于规模较大的 OpenStack 环境。

控制节点部署 neutron-server 服务，只负责通过 neutron-server 响应 API 请求。

网络节点部署的服务包括 Core Plugin 的代理、Service Plugin 的代理。可将所有的代理主键从上述控制节点分离出来，部署到独立的网络节点上，由独立的网络节点实现数据的交换、路由以及负载均衡等高级网络服务。

5.2.3　Neutron 主要插件、代理与服务

Neutron 主要插件、代理和服务的层次结构如图 5.8 所示。

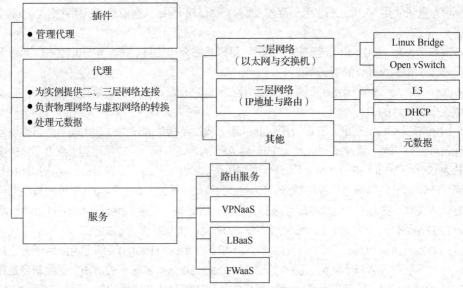

图 5.8　Neutron 主要插件、代理和服务的层次结构

1. 多种二层网络技术插件

Neutron 可以通过开发不同的插件和代理来支持不同的网络技术，这是一种相当开放的架构。但随着所支持的网络提供者的增加，开发人员发现了两个突出的问题。一个问题是多种网络提供者无法共存。Core Plugin 负责管理和维护 Neutron 二层虚拟网络的状态信息，一个 Neutron 网络只能由一个插件管理，而 Core Plugin 与相应的代理是一一对应的。如果选择 Linux Bridge 插件，则只能由 Linux Bridge 代理，且必须在 OpenStack 的所有节点上使用 Linux Bridge 作为虚拟交换机。另一个问题是开发新的插件的工作量太大，所有传统的 Core Plugin 中存在大量重复代码，如数据库访问代码等。

为解决这两个问题，从 OpenStack 的 Havana 版本开始，Neutron 实现了一个多种模块层 2（Module Layer 2，ML2）插件，旨在取代所有的 Core Plugin，允许在 OpenStack 网络中同时使用多种二层网络技术，不同的节点可以使用不同的网络实现机制。ML2 插件能够与现有的代理无缝集成，以前使用的代理无须变更，只需将 Core Plugin 替换为 ML2 插件即可。ML2 插件对新的网络技术的支持更为简单，无须从头开发 Core Plugin，只需开发相应的机制驱动，可大大减少要编写和维护的代码。

ML2 插件架构如图 5.9 所示，ML2 插件对二层网络进行抽象，解耦 Neutron 所支持的网络类型与访问这些网络类型的虚拟网络实现机制，并通过驱动的形式进行扩展。不同的网络类型对应类型驱动，由类型管理器进行管理。不同的网络实现机制对应机制驱动，由机制管理器进行管理。这种实现架构使 ML2 插件具有弹性，易于扩展，能够灵活支持多种网络类型和实现机制。

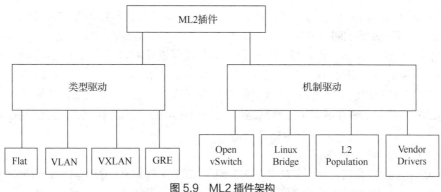

图 5.9　ML2 插件架构

（1）类型驱动

Neutron 支持的每一种网络类型都有一个对应的 ML2 插件类型驱动。类型驱动负责维护网络类型的状态，并执行验证、创建网络等操作。目前，Neutron 已经实现的网络类型包括 Flat、VLAN、VXLAN 和 GRE 等。

（2）机制驱动

机制驱动负责获取由类型驱动维护的网络状态，并确保在相应网络设备（物理的或虚拟的）上正确实现这些状态。例如，类型驱动为 VLAN，机制驱动为 Linux Bridge。如果创建网络 VLAN 10，那么 VLAN 类型驱动会确保将 VLAN 10 的信息保存在 Neutron 数据库中，包括网络的名称化 VLAN ID 等。而 Linux Bridge 代理在物理网卡上创建 ID 为 10 的 VLAN 设备和 Bridge 设备，并将两者进行桥接。

（3）扩展资源

ML2 插件作为一个 Core Plugin，在实现网络、子网和端口核心资源的同时，也实现了包括端口绑定、安全组等部分扩展资源。总之，ML2 插件已经成为 Neutron 的首选插件。

2. Linux Bridge 代理

Linux Bridge 是成熟可靠的 Neutron 二层网络虚拟化技术，支持 Local、Flat、VLAN 和 VXLAN 这 4 种网络类型，目前不支持 GRE。

Linux Bridge 可以将一台主机上的多块网卡桥接起来，充当一台交换机，它既可以桥接物理网卡，又可以桥接虚拟网卡。用于桥接虚拟网卡的是 Tap 接口，这是一个虚拟出来的网络设备，称为 Tap 设备。作为网桥的一个端口，Tap 接口在逻辑上与物理接口具有相同的功能，可以接收和发送数据报文。

3. Open vSwitch 代理

与 Linux Bridge 代理相比，Open vSwitch 具有集中管控功能，而且性能更加优化，支持更多的功能，目前在 OpenStack 领域中成为主流。它支持 Local、Flat、VLAN、VXLAN、GRE 和 GENEVE 等网络类型。

4. DHCP 代理

OpenStack 的实例在启动过程中能够从 Neutron 提供的 DHCP 服务中自动获得 IP 地址。

（1）DHCP 代理（neutron-dhcp-agent）：为项目网络提供 DHCP 功能，提供元数据请求（Metadata Request）服务。

（2）DHCP 驱动：用于管理 DHCP 服务器，默认为 dnsmasq。这是一款提供 DHCP 和 DNS 服务的开源软件，可提供 DNS 缓存和 DHCP 服务功能。

（3）DHCP 代理调度器（Agent Scheduler）：负责 DHCP 代理与网络（Neutron）的调度。

5. Linux 网络命名空间

Linux 网络命名空间是 Linux 提供的一种内核级别网络环境隔离的方法，当前 Linux 支持不同类型的命名空间，网络命名空间只是其中的一种。在二层网络上，VLAN 可以将一个物理交换机分割成几个独立的虚拟交换机。类似的，在三层网络上，Linux 网络命名空间可以将一个物理三层网络分割成几个独立的虚拟三层网络，作为一种资源虚拟隔离机制，它在 Neutron 中得以应用。

（1）Linux 网络命名空间简介

在 Linux 中，网络命名空间可以被认为是隔离的、拥有单独网络栈的环境，如网络接口、路由、iptables 等。它经常用来隔离网络资源，如设备、服务，只有拥有同样网络命名空间的设备才能彼此访问；它还提供了在网络命名空间内运行进程的功能。后台进程可以运行在不同网络命名空间内的相同端口上，用户还可以虚拟出一块网卡。

网络命名空间可以创建一个完全隔离的全新网络环境，包括一个独立的网络接口、路由表、ARP 表、IP 地址表、iptables 或 ebtables 等。与网络有关的组件都是独立的。

通常情况下，可以使用 ip netns add 命令添加新的网络命名空间，使用 ip netns list 命令查看所有的网络命名空间。

网络命名空间内部的通信没有问题，但被隔离的网络命名空间之间要进行通信，就必须采用特定方法，即虚拟网络（Virtual Ethernet，VETH）接口对。VETH 接口对是一种成对出现的特殊网络设备，它们像一根虚拟的网线，可用于连接两个网络命名空间，向 VETH 接口对的一端输入的数据将自动转发到另一端。

创建一对 VETH 接口类型管道（Pipe），发送给虚拟网卡 1（veth1）的数据报文可以在虚拟网卡 2（veth2）上收到，发送给 veth2 的数据报文也可以在 veth1 上收到，相当于安装两个接口并用网线连接起来。将上述两个 VETH 接口分别放置到两个网络命名空间中，这样两个 VETH 接口就分别出现在两个网络命名空间中，两个空间就打通了，其中的设备可以相互访问。

（2）Linux 网络命名空间实现 DHCP 服务隔离

Neutron 通过网络命名空间为每个网络提供独立的 DHCP 和路由服务，从而允许项目创建重叠的网络。如果没有这种隔离机制，网络就不能重叠，这样就失去了很多灵活性。

dnsmasq 是可以方便地配置 DNS 和 DHCP 的工具，适用于小型网络，提供了 DNS 功能和可选择的 DHCP 功能。每个 dnsmasq 进程都位于独立的网络命名空间，其名称为 qdhcp-xx。

例如，以创建 Flat 网络为例，Neutron 自动新建该网络对应的网桥 brq-xx，以及 DHCP 的 Tap 设备 tap-xx。物理主机也有一个网络命名空间，称为 root，拥有所有物理和虚拟接口设备，而物理接口只能位于 root 网络命名空间。新创建的网络命名空间默认只有一个回环设备（Loopback Device）。如果 DHCP 的 Tap 虚拟接口放置到 qdhcp-xx 网络命名空间中，则该 Tap 虚拟接口将无法直接与 root 网络命名空间中的网桥设备 brq-xx 连接。为此，Neutron 使用 VETH 接口对来解决这个问题，添加 VETH 接口对 tap-xx 与 ns-xx，让 qdhcp-xx 连接到 brq-xx。

（3）Linux 网络命名空间实现路由器

Neutron 允许在不同网络中的子网的无类别域间路由选择（Classless Inter-Domain Routing，CIDR）和 IP 地址重叠，具有相同 IP 地址的两个虚拟机也不会产生冲突。这是由 Neutron 的路由器通过 Linux 网络命名空间实现的，每个路由器都有自己独立的路由表。

6. Neutron 路由器

Neutron 路由器是一个三层网络的抽象，其模拟物理路由器，为用户提供路由、NAT 等服务。

在 OpenStack 网络中，不同子网之间的通信需要通过路由器实现，项目网络与外部网络之间的通信更需要通过路由器实现。

Neutron 提供了虚拟路由器，也支持物理路由器。例如，两个隔离的 VLAN 要进行通信，可以通过物理路由器实现，如图 5.10 所示，由物理路由器提供相应的 IP 路由表，确保两个 IP 子网之间的通信。将两个 VLAN 中的虚拟机的默认网关分别设置为物理路由器的接口 A 和 B 的 IP 地址，当 VLAN 10 中的虚拟机要与 VLAN 20 中的虚拟机通信时，数据报文将通过 VLAN 10 中的物理网卡到达物理路由器，由物理路由器转到 VLAN 20 中的物理网卡，再转到目的虚拟机。

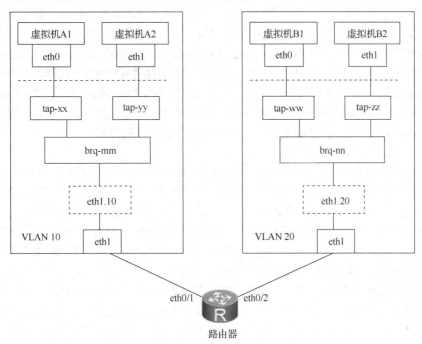

图 5.10　通过 Neutron 路由器实现通信

Neutron 的虚拟路由器使用软件模拟物理路由器，路由实现机制相同。Neutron 的路由服务由代理提供实现。

7. L3 代理

在 Neutron 中，L3（neutron-l3-agent，L3）代理具有十分重要的地位，它不仅提供虚拟路由器，还通过 iptables 提供地址转换、浮动地址（Floating IP）和安全组（Security Group）功能。L3 代理利用 Linux IP 栈、路由和 iptables 来实现内部网络中不同网络的虚拟机实例之间的通信，以及虚拟机实例和外部网络之间的网络流量的路由和转发。L3 代理可以部署在控制节点或者网络节点上。

（1）路由

L3 代理提供的虚拟路由器通过虚拟接口连接到子网，一个子网一个接口，该接口的地址是该子网的网关地址。如果虚拟机的 IP 栈发现数据报文的目的 IP 地址不在本网段，则会将其发到路由器上对应其子网的虚拟接口。此后，虚拟机路由器根据配置的路由规则和目的 IP 地址将数据报文转发到目的端口。L3 代理会为每个路由器创建一个网络命名空间，通过 VETH 接口对与中断线适配器（Trunk AdaPter，TAP）相连，并将网关 IP 地址配置在位于网络命名空间的 VETH 接口上，这样就能够提供路由。如果网络节点不支持 Linux 网络命名空间，则只能运行一个虚拟路由器。

（2）通过网络命名空间支持网络重叠

在云计算环境下，用户可以按照自己的规划创建网络，不同项目的网络 IP 地址可能会重叠。为实现此功能，L3 代理使用 Linux 网络命名空间来提供隔离的转发上下文，隔离不同项目的网络。每个 L3 代理运行在一个地址命名空间中，每个地址命名空间由 qrouter-<router-UUID> 命名。

（3）源地址转换

L3 代理通过在 iptables 表中增加 Postrouting 链来实现源地址转换（Source Network Address Translation，SNAT），即内网计算机访问外网时，发起访问的内网 IP 地址（源 IP 地址）转换为外网网关的 IP 地址。这种功能让虚拟机实例能够直接访问外网。但外网计算机还不能直接访问虚拟机实例，因为实例没有外网 IP 地址，而目的地址转换就能解决这一问题。项目网络连接到 Neutron 路由器时，通常将路由器作为默认网关。当路由器接收到实例的数据报文并将其转发到外网时，路由器会将数据报文的源地址修改成自己的外网地址，确保数据报文转发到外网，并能够从外网返回。路由器修改返回的数据报文，并转发给之前发起访问的实例。

（4）目的地址转换与浮动 IP 地址

Neutron 需要设置浮动 IP 地址支持从外网访问项目网络中的实例。每个浮动 IP 地址对应一个唯一的路由器。浮动 IP 地址→关联的端口→所在的子网→包含该子网以及外部子网的路由器。创建浮动 IP 地址时，在 Neutron 分配浮动 IP 地址后，通过 RPC 通知该浮动 IP 地址对应的路由器去设置该浮动 IP 地址对应的 iptables 规则。从外网访问虚拟机实例时，目的 IP 地址为实例的浮动 IP 地址，因此必须由 iptables 将其转换为固定 IP 地址，再将其路由到实例。L3 代理通过在 iptables 表中增加 Preouting 链来实现目的地址转换（Destination Network Address Translation，DNAT）。

浮动 IP 地址提供静态 NAT 功能，建立外网 IP 地址与实例所在项目 IP 地址的一对一映射。浮动 IP 地址配置在路由器提供网关的外网接口上，而不是实例中。路由器会根据通信的方向修改数据报文的源地址或者目的地址，这是通过在路由器上应用 iptables 的 NAT 规则来实现的。一旦设置浮动 IP 地址后，源地址转换就不再使用外网网关的 IP 地址了，而是直接使用对应的浮动 IP 地址。虽然相关的 NAT 规则依然存在，但是 neutron-l3-agent-float-snat 比 neutron-l3-agent-snat 更早执行。

（5）安全组

安全组（Security Group）定义了哪些进入的网络流量能被转发给虚拟机实例。安全组包含一组防火墙策略，称为安全组规则（Security Group Rule）。用户可以定义若干个安全组，每个安全组可以有若干条规则，每个实例可以绑定若干个安全组。安全组通过 iptables 对实例所在计算节点的网络流量进行过滤。安全组规则作用在实例的端口上，具体是在连接实例的计算节点上的 Linux 网桥上实施。

8. 虚拟防火墙

防火墙即服务（FireWall-as-a-Service，FWaaS）是一种基于 L3 代理的防火墙服务，是 Neutron 的一个高级服务。通过使用 FWaaS，OpenStack 可以将防火墙应用到项目、路由器端口和虚拟机端口上，在子网边界上对三层和四层的流量进行过滤。

传统网络中的防火墙一般放在网关上，用来控制子网之间的访问。FWaaS 也一样，在 Neutron 路由器上应用防火墙规则，控制进出项目网络的数据。防火墙必须关联某个策略。策略是规则的集合，防火墙会按顺序应用策略中的每一条规则。规则是访问控制的规则，由源与目的子网 IP 地址、源与目的端口、协议、允许或拒绝动作组成。

安全组是最早的网络安全模块，其应用对象是虚拟路由器，可以在安全组之前控制从外部传入的流量，但是对于同一个子网内的流量不做限制。安全组保护的是实例，而 FWaaS 保护的是子网，两者互为补充，通常同时部署 FWaaS 和安全组来实现双重防护。

FWaaS v1 是传统的防火墙方案，用于对路由器提供保护功能。将虚拟防火墙应用到路由器时，该路由器的所有内部端口都会受到保护，如图 5.11 所示。

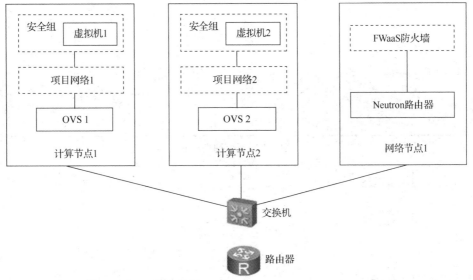

图 5.11　虚拟防火墙的应用

FWaaS v2 提供了更具细粒度的安全服务。防火墙由防火墙组（Firewall Group）替代，一个防火墙一般包括两项策略：入口策略（Ingress Policy）和出口策略（Egress Policy）。防火墙组不再应用于路由器级，而应用于路由器的全部端口。注意，FWaaS v2 的配置仅提供命令行工具，不支持 Dashboard 界面。

5.3　项目实施

5.3.1　基于 Web 界面管理网络服务

基于部署远程桌面管理 OpenStack 平台，可以验证和操作网络服务。用户以云管理员的身份登录 Dashboard 界面，可以执行网络服务管理操作。

1. 查看网络列表信息

在 Dashboard 界面中，单击"管理员"主节点，再单击"系统"→"网络"子节点，弹出网络列表界面，如图 5.12 所示，可以查看网络列表信息。

2. 创建网络

（1）创建内部网络

在网络列表界面的右上方单击"+创建网络"按钮，打开"创建网络"对话框，如图 5.13 所示。

根据需要创建新网络，可以创建供应商规定的网络，可以为新的虚拟网络指定物理网络类型（如 Flat、VLAN、GRE、和 VXLAN）及其段 ID（segmentation_id），或者物理网络名称。

此外，可以选中相应的复选框来创建外部网络或者共享网络。

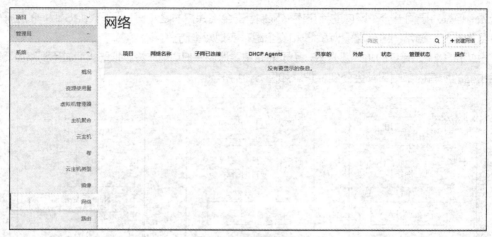

图 5.12 网络列表界面

图 5.13 "创建网络"对话框

这里创建内部网络，选择相应的网络类型并指定网络的名称，单击"提交"按钮，返回网络列表界面，可以查看网络的列表信息，如图 5.14 所示。

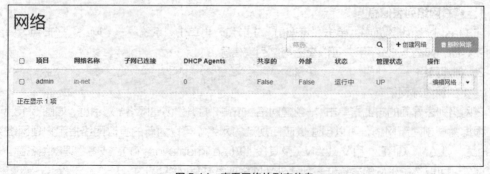

图 5.14 查看网络的列表信息

在网络列表界面中，选中"网络名称"下的内网名称"in-net"，弹出网络概况界面，可以查看内部网络概况，如图 5.15 所示。

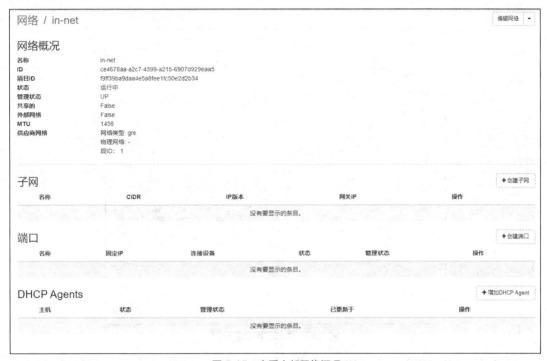

图 5.15　查看内部网络概况

在网络概况界面的右侧单击"+创建子网"按钮，打开"创建子网"对话框，如图 5.16 所示，创建内部子网，可以进行子网名称、网络地址、IP 版本、网关 IP 以及禁用网关等相关设置。

图 5.16　"创建子网"对话框

选择"子网详情"选项卡，可进行内部子网高级配置，如图 5.17 所示。在"子网详情"选项卡中，可以选中"激活 DHCP"复选框，为子网指定扩展属性，如分配地址池、DNS 服务器、主机路由等。

图 5.17　内部子网高级配置

单击"已创建"按钮，返回网络概况界面，可以查看子网、端口、DHCP Agents 等关联内部子网详情，如图 5.18 所示。

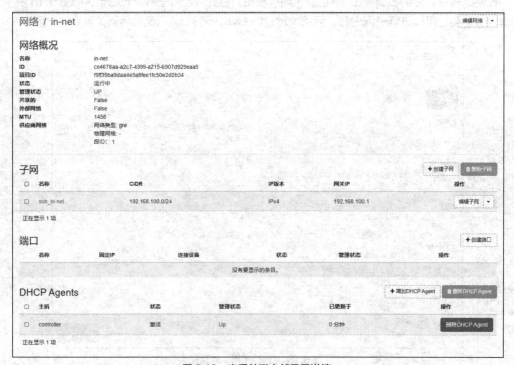

图 5.18　查看关联内部子网详情

在 Dashboard 界面中，单击"管理员"主节点，再单击"系统"→"网络"子节点，弹出网络列表界面，查看当前网络关联内部子网列表情况，如图 5.19 所示。

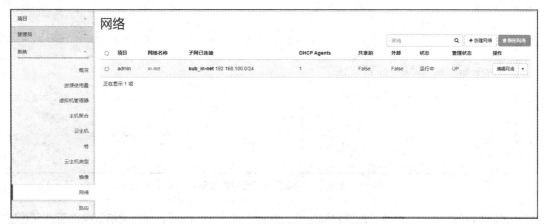

图 5.19　查看当前网络关联内部子网列表情况

（2）创建外部网络

单击"管理员"主节点，再单击"系统"→"网络"子节点，弹出网络列表界面。在网络列表界面的右侧单击"+创建网络"按钮，打开"创建网络"对话框，如图 5.20 所示。

图 5.20　创建外部网络

这里创建外部网络，选择相应的网络类型并指定网络的名称，选中"外部网络"复选框，单击"提交"按钮，返回网络列表界面，可以查看外部网络列表信息，如图 5.21 所示。

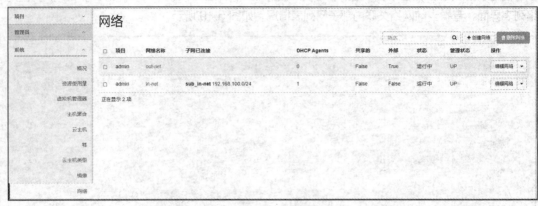

图 5.21　查看外部网络列表信息

在网络列表界面中，选中"网络名称"下的外网名称"out-net"，弹出网络概况界面，可以查看外部网络概况，如图 5.22 所示。

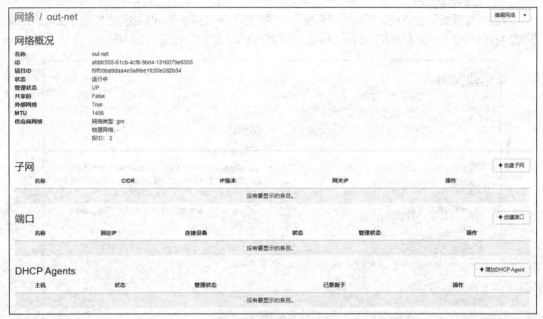

图 5.22　查看外部网络概况

在网络概况界面的右侧单击"+创建子网"按钮，打开"创建子网"对话框，创建外部子网，如图 5.23 所示，可以进行子网名称、网络地址、IP 版本、网关 IP 以及禁用网关等相关设置。

创建关联到这个网络的子网，选择"子网详情"选项卡，可进行外部子网高级配置，如图 5.24 所示。在子网详情界面中可以选中"激活 DHCP"复选框，为子网指定扩展属性，如分配地址池、DNS 服务器、主机路由等。

单击"已创建"按钮，返回网络概况界面，可以查看子网、端口、DHCP Agents 等相关信息，如图 5.25 所示。

图 5.23 创建外部子网

图 5.24 外部子网高级配置

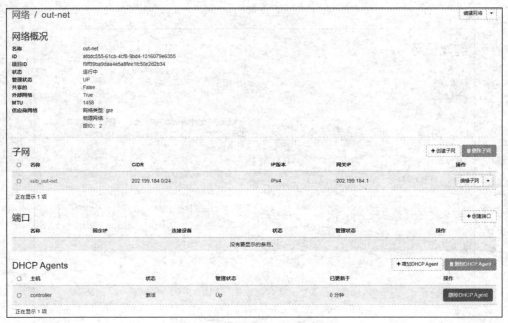

图 5.25　查看关联外部子网详情

单击"管理员"主节点，再单击"系统"→"网络"子节点，弹出网络列表界面，查看当前网络关联外部子网列表情况，如图 5.26 所示。

图 5.26　查看当前网络关联外部子网列表情况

3. 添加路由

在 Dashboard 界面中，单击"项目"主节点，再单击"网络"→"路由"子节点，弹出路由列表界面，可以查看当前路由列表情况，如图 5.27 所示。

图 5.27　查看当前路由列表情况

在路由列表界面右上方单击"+新建路由"按钮,打开"新建路由"对话框,如图 5.28 所示。

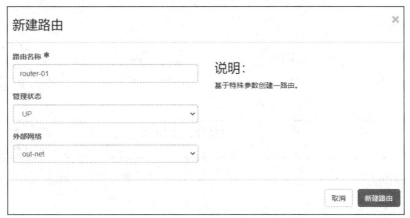

图 5.28 "新建路由"对话框

在"新建路由"对话框中,指定路由名称、管理状态、外部网络,单击"新建路由"按钮,返回路由列表界面,即可查看新建路由列表信息,如图 5.29 所示。

图 5.29 查看新建路由列表信息

在路由列表界面中,选中"名称"下的路由名称"router-01",可以查看路由信息,如图 5.30 所示。

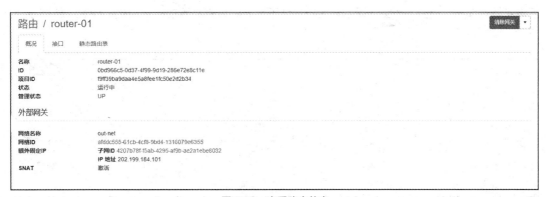

图 5.30 查看路由信息

在路由信息界面中,选择"接口"选项卡,可以查看路由接口信息,如图 5.31 所示。

在"接口"选项卡中,单击右侧的"增加接口"按钮,打开"增加接口"对话框,如图 5.32 所示。

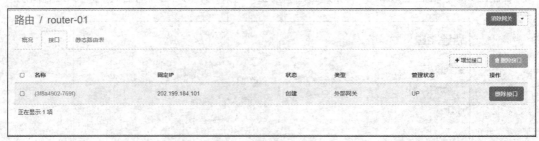

图 5.31 查看路由接口信息

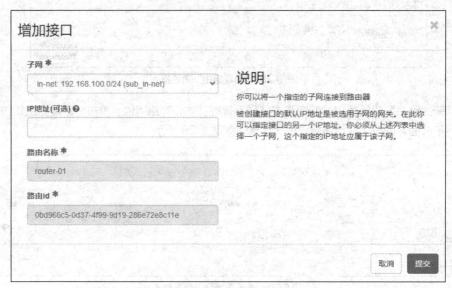

图 5.32 "增加接口"对话框

可以将一个指定的子网连接到路由器上，被创建接口的默认 IP 地址是被选用子网的网关。在此可以指定接口的另一个 IP 地址。必须从上述列表中选择一个子网，这个指定的 IP 地址应属于该子网。单击"提交"按钮，返回"接口"选项卡，即可查看添加的路由接口信息，如图 5.33 所示。

图 5.33 查看添加的路由接口信息

在 Dashboard 界面中，单击"项目"主节点，再单击"网络"→"网络拓扑"子节点，弹出网络拓扑界面，可以查看当前网络拓扑情况，如图 5.34 所示。

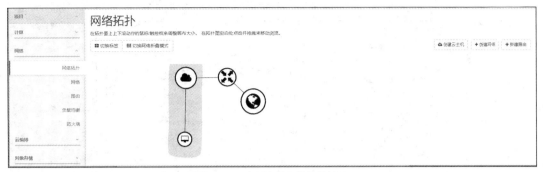

图 5.34 查看当前网络拓扑情况

5.3.2 基于命令行界面管理网络服务

系统管理员一般会基于命令行界面管理网络服务，建议使用 openstack 命令来代替 neutron 命令。

1. 查看网络信息

查看网络列表信息，可以使用 openstack 命令，也可以使用 neutron 命令，执行命令如下。

V5-4 查看网络信息

```
#openstack  network  list
```
或者
```
#neutron  net-list
```
其命令的结果如图 5.35 所示。

```
[root@controller ~]# openstack network list
+--------------------------------------+---------+--------------------------------------+
| ID                                   | Name    | Subnets                              |
+--------------------------------------+---------+--------------------------------------+
| ce4678aa-a2c7-4399-a215-6907d929eaa5 | in-net  | 83cb607b-8e56-4d8f-b7b5-d84c375d660a |
| afddc555-61cb-4cf8-9bd4-1316079e6355 | out-net | 4207b78f-f5ab-4295-af9b-ae2a1ebe8032 |
+--------------------------------------+---------+--------------------------------------+
[root@controller ~]# neutron net-list
+--------------------------------------+---------+-------------------------------------------------------+
| id                                   | name    | subnets                                               |
+--------------------------------------+---------+-------------------------------------------------------+
| ce4678aa-a2c7-4399-a215-6907d929eaa5 | in-net  | 83cb607b-8e56-4d8f-b7b5-d84c375d660a 192.168.100.0/24 |
| afddc555-61cb-4cf8-9bd4-1316079e6355 | out-net | 4207b78f-f5ab-4295-af9b-ae2a1ebe8032 202.199.184.0/24 |
+--------------------------------------+---------+-------------------------------------------------------+
[root@controller ~]#
```

图 5.35 查看网络列表信息

查看某个网络的详细信息，命令格式如下。
```
#openstack  network  show  网络名称或ID
```
或者
```
#neutron  net-show  网络名称或ID
```
例如，查看内部网络 in-net 的详细信息，其命令的结果如图 5.36 所示。

查看子网列表信息，执行命令如下。
```
#openstack  subnet  list
```
或者
```
#neutron  subnet-list
```
其命令的结果如图 5.37 所示。

```
[root@controller ~]# openstack network show in-net
+---------------------------+--------------------------------------+
| Field                     | Value                                |
+---------------------------+--------------------------------------+
| admin_state_up            | UP                                   |
| availability_zone_hints   |                                      |
| availability_zones        | nova                                 |
| created_at                | 2021-01-10T20:58:54                  |
| description               |                                      |
| id                        | ce4678aa-a2c7-4399-a215-6907d929eaa5 |
| ipv4_address_scope        | None                                 |
| ipv6_address_scope        | None                                 |
| mtu                       | 1458                                 |
| name                      | in-net                               |
| port_security_enabled     | True                                 |
| project_id                | f9ff39ba9daa4e5a8fee1fc50e2d2b34     |
| provider:network_type     | gre                                  |
| provider:physical_network | None                                 |
| provider:segmentation_id  | 1                                    |
| router_external           | Internal                             |
| shared                    | False                                |
| status                    | ACTIVE                               |
| subnets                   | 83cb607b-8e56-4d8f-b7b5-d84c375d660a |
| tags                      | []                                   |
| updated_at                | 2021-01-10T20:58:54                  |
+---------------------------+--------------------------------------+
[root@controller ~]# neutron net-show in-net
+---------------------------+--------------------------------------+
| Field                     | Value                                |
+---------------------------+--------------------------------------+
| admin_state_up            | True                                 |
| availability_zone_hints   |                                      |
| availability_zones        | nova                                 |
| created_at                | 2021-01-10T20:58:54                  |
| description               |                                      |
| id                        | ce4678aa-a2c7-4399-a215-6907d929eaa5 |
| ipv4_address_scope        |                                      |
| ipv6_address_scope        |                                      |
| mtu                       | 1458                                 |
| name                      | in-net                               |
| port_security_enabled     | True                                 |
| provider:network_type     | gre                                  |
| provider:physical_network |                                      |
| provider:segmentation_id  | 1                                    |
| router:external           | False                                |
| shared                    | False                                |
| status                    | ACTIVE                               |
| subnets                   | 83cb607b-8e56-4d8f-b7b5-d84c375d660a |
| tags                      |                                      |
| tenant_id                 | f9ff39ba9daa4e5a8fee1fc50e2d2b34     |
| updated_at                | 2021-01-10T20:58:54                  |
+---------------------------+--------------------------------------+
[root@controller ~]#
```

图 5.36 查看内部网络 in-net 的详细信息

```
[root@controller ~]# openstack subnet list
+--------------------------------------+-------------+--------------------------------------+------------------+
| ID                                   | Name        | Network                              | Subnet           |
+--------------------------------------+-------------+--------------------------------------+------------------+
| 83cb607b-8e56-4d8f-b7b5-d84c375d660a | sub_in-net  | ce4678aa-a2c7-4399-a215-6907d929eaa5 | 192.168.100.0/24 |
| 4207b78f-f5ab-4295-af9b-ae2a1ebe8032 | sub_out-net | afddc555-61cb-4cf8-9bd4-1316079e6355 | 202.199.184.0/24 |
+--------------------------------------+-------------+--------------------------------------+------------------+
[root@controller ~]# neutron subnet-list
+--------------------------------------+-------------+------------------+--------------------------------------------------------+
| id                                   | name        | cidr             | allocation_pools                                       |
+--------------------------------------+-------------+------------------+--------------------------------------------------------+
| 83cb607b-8e56-4d8f-b7b5-d84c375d660a | sub_in-net  | 192.168.100.0/24 | {"start": "192.168.100.100", "end": "192.168.100.200"} |
| 4207b78f-f5ab-4295-af9b-ae2a1ebe8032 | sub_out-net | 202.199.184.0/24 | {"start": "202.199.184.100", "end": "202.199.184.200"} |
+--------------------------------------+-------------+------------------+--------------------------------------------------------+
[root@controller ~]#
```

图 5.37 查看子网列表信息

查看某个子网的详细信息，命令格式如下。

#openstack subnet show 网络名称或 ID

或者

#neutron subnet-show   网络名称或 ID

例如，查看内部子网 sub_in-net 的详细信息，其命令的结果如图 5.38 所示。

查看路由列表信息，执行命令如下。

#openstack router list

或者

neutron router-list

其命令的结果如图 5.39 所示。

```
[root@controller ~]# openstack subnet show sub_in-net
+-------------------+------------------------------------------------+
| Field             | Value                                          |
+-------------------+------------------------------------------------+
| allocation_pools  | 192.168.100.100-192.168.100.200                |
| cidr              | 192.168.100.0/24                               |
| created_at        | 2021-01-10T21:05:42                            |
| description       |                                                |
| dns_nameservers   |                                                |
| enable_dhcp       | True                                           |
| gateway_ip        | 192.168.100.1                                  |
| host_routes       |                                                |
| id                | 83cb607b-8e56-4d8f-b7b5-d84c375d660a           |
| ip_version        | 4                                              |
| ipv6_address_mode | None                                           |
| ipv6_ra_mode      | None                                           |
| name              | sub_in-net                                     |
| network_id        | ce4678aa-a2c7-4399-a215-6907d929eaa5           |
| project_id        | f9ff39ba9daa4e5a8fee1fc50e2d2b34               |
| subnetpool_id     | None                                           |
| updated_at        | 2021-01-10T21:05:42                            |
+-------------------+------------------------------------------------+
[root@controller ~]# neutron subnet-show sub_in-net
+-------------------+------------------------------------------------------+
| Field             | Value                                                |
+-------------------+------------------------------------------------------+
| allocation_pools  | {"start": "192.168.100.100", "end": "192.168.100.200"}|
| cidr              | 192.168.100.0/24                                     |
| created_at        | 2021-01-10T21:05:42                                  |
| description       |                                                      |
| dns_nameservers   |                                                      |
| enable_dhcp       | True                                                 |
| gateway_ip        | 192.168.100.1                                        |
| host_routes       |                                                      |
| id                | 83cb607b-8e56-4d8f-b7b5-d84c375d660a                 |
| ip_version        | 4                                                    |
| ipv6_address_mode |                                                      |
| ipv6_ra_mode      |                                                      |
| name              | sub_in-net                                           |
| network_id        | ce4678aa-a2c7-4399-a215-6907d929eaa5                 |
| subnetpool_id     |                                                      |
| tenant_id         | f9ff39ba9daa4e5a8fee1fc50e2d2b34                     |
| updated_at        | 2021-01-10T21:05:42                                  |
+-------------------+------------------------------------------------------+
[root@controller ~]#
```

图 5.38 查看内部子网 sub_in-net 的详细信息

```
[root@controller ~]# openstack router list
+--------------------------------------+-----------+--------+-------+-------------+-------+----------------------------------+
| ID                                   | Name      | Status | State | Distributed | HA    | Project                          |
+--------------------------------------+-----------+--------+-------+-------------+-------+----------------------------------+
| 0bd966c5-0d37-4f99-9d19-286e72e8c11e | router-01 | ACTIVE | UP    | False       | False | f9ff39ba9daa4e5a8fee1fc50e2d2b34 |
+--------------------------------------+-----------+--------+-------+-------------+-------+----------------------------------+
[root@controller ~]# neutron router-list
+--------------------------------------+-----------+---------------------------------------------------------+-------------+-------+
| id                                   | name      | external_gateway_info                                   | distributed | ha    |
+--------------------------------------+-----------+---------------------------------------------------------+-------------+-------+
| 0bd966c5-0d37-4f99-9d19-286e72e8c11e | router-01 | {"network_id": "afddc555-61cb-4cf8-9bd4-1316079e6355",  | False       | False |
|                                      |           | "enable_snat": true, "external_fixed_ips": [{"subnet_id":|            |       |
|                                      |           | "4207b78f-f5ab-4295-af9b-ae2a1ebe8032", "ip_address":   |             |       |
|                                      |           | "202.199.184.101"}]}                                    |             |       |
+--------------------------------------+-----------+---------------------------------------------------------+-------------+-------+
[root@controller ~]#
```

图 5.39 查看路由列表信息

查看路由详细信息,命令格式如下。

#openstack router show 路由名称或 ID

或者

neutron router-show 路由名称或 ID

其命令的结果如图 5.40 所示。

```
[root@controller ~]# openstack router show router-01
+-------------------------+-------------------------------------------------------------------------------+
| Field                   | Value                                                                         |
+-------------------------+-------------------------------------------------------------------------------+
| admin_state_up          | UP                                                                            |
| availability_zone_hints |                                                                               |
| availability_zones      | nova                                                                          |
| description             |                                                                               |
| distributed             | False                                                                         |
| external_gateway_info   | {"network_id": "afddc555-61cb-4cf8-9bd4-1316079e6355", "enable_snat": true,   |
|                         | "external_fixed_ips": [{"subnet_id": "4207b78f-                                |
|                         | f5ab-4295-af9b-ae2a1ebe8032", "ip_address": "202.199.184.101"}]}              |
| ha                      | False                                                                         |
| id                      | 0bd966c5-0d37-4f99-9d19-286e72e8c11e                                          |
| name                    | router-01                                                                     |
| routes                  | []                                                                            |
| status                  | ACTIVE                                                                        |
| tenant_id               | f9ff39ba9daa4e5a8fee1fc50e2d2b34                                              |
+-------------------------+-------------------------------------------------------------------------------+
[root@controller ~]# neutron router-show router-01
+-------------------------+-------------------------------------------------------------------------------+
| Field                   | value                                                                         |
+-------------------------+-------------------------------------------------------------------------------+
| admin_state_up          | True                                                                          |
| availability_zone_hints |                                                                               |
| availability_zones      | nova                                                                          |
| description             |                                                                               |
| distributed             | False                                                                         |
| external_gateway_info   | {"network_id": "afddc555-61cb-4cf8-9bd4-1316079e6355", "enable_snat": true,   |
|                         | "external_fixed_ips": [{"subnet_id": "4207b78f-                                |
|                         | f5ab-4295-af9b-ae2a1ebe8032", "ip_address": "202.199.184.101"}]}              |
| ha                      | False                                                                         |
| id                      | 0bd966c5-0d37-4f99-9d19-286e72e8c11e                                          |
| name                    | router-01                                                                     |
| routes                  |                                                                               |
| status                  | ACTIVE                                                                        |
| tenant_id               | f9ff39ba9daa4e5a8fee1fc50e2d2b34                                              |
+-------------------------+-------------------------------------------------------------------------------+
[root@controller ~]#
```

图 5.40 查看路由详细信息

查看网络服务列表信息，执行命令如下。

```
# neutron  agent-list
```

其命令的结果如图 5.41 所示。

```
[root@controller ~]# neutron agent-list
+--------------------------------------+------------------+------------+-------------------+-------+----------------+---------------------------+
| id                                   | agent_type       | host       | availability_zone | alive | admin_state_up | binary                    |
+--------------------------------------+------------------+------------+-------------------+-------+----------------+---------------------------+
| 1ccc8d2d-b40a-4406-b15b-ab4b2b8e8398 | Open vSwitch agent | controller |                   | :-)   | True           | neutron-openvswitch-agent |
| 245ae2f9-6220-4c00-a36d-542a746a3f9e | DHCP agent       | controller | nova              | :-)   | True           | neutron-dhcp-agent        |
| 26f38d05-3553-46f6-8740-847f97741f89 | L3 agent         | controller | nova              | :-)   | True           | neutron-l3-agent          |
| 5bd0c9e2-a2ef-4890-a4ac-a4cb6bccf71b | Metadata agent   | controller |                   | :-)   | True           | neutron-metadata-agent    |
| f1d1fc34-6e9d-46b5-8370-0579f3b46b6f | Loadbalancer agent | controller |                 | :-)   | True           | neutron-lbaas-agent       |
+--------------------------------------+------------------+------------+-------------------+-------+----------------+---------------------------+
[root@controller ~]#
```

图 5.41 查看网络服务列表信息

使用 neutron 命令查看网络服务列表信息中某一列的信息，命令格式如下。

```
# neutron  agent-list  -c  列名
```

例如，查看网络服务列表中 agent_type 列的信息，其命令的结果如图 5.42 所示。

```
[root@controller ~]# neutron agent-list -c agent_type
+--------------------+
| agent_type         |
+--------------------+
| Open vSwitch agent |
| DHCP agent         |
| L3 agent           |
| Metadata agent     |
| Loadbalancer agent |
+--------------------+
[root@controller ~]#
```

图 5.42 查看网络服务列表中 agent_type 列的信息

例如，查看网络服务列表中 id 与 agent_type 列的信息，命令格式如下。

```
# neutron  agent-list  -c  列名 1  -c  列名 2
```

其命令的结果如图 5.43 所示。

```
[root@controller ~]# neutron agent-list -c id -c agent_type
+--------------------------------------+--------------------+
| id                                   | agent_type         |
+--------------------------------------+--------------------+
| 1ccc8d2d-b40a-4406-b15b-ab4b2b8e8398 | Open vSwitch agent |
| 245ae2f9-6220-4c00-a36d-542a746a3f9e | DHCP agent         |
| 26f38d05-3553-46f6-8740-847f97741f89 | L3 agent           |
| 5bd0c9e2-a2ef-4890-a4ac-a4cb6bccf71b | Metadata agent     |
| f1d1fc34-6e9d-46b5-8370-0579f3b46b6f | Loadbalancer agent |
+--------------------------------------+--------------------+
[root@controller ~]#
```

图 5.43 查看网络服务列表中 id 与 agent_type 列的信息

查看网络服务列表某个服务的详细信息，命令格式如下。

```
# neutron agent-show 服务 ID
```

例如，查看 L3 agent 服务的详细信息，其命令的结果如图 5.44 所示。

```
[root@controller ~]# neutron agent-show 26f38d05-3553-46f6-8740-847f97741f89
+---------------------+----------------------------------------------------------+
| Field               | Value                                                    |
+---------------------+----------------------------------------------------------+
| admin_state_up      | True                                                     |
| agent_type          | L3 agent                                                 |
| alive               | True                                                     |
| availability_zone   | nova                                                     |
| binary              | neutron-l3-agent                                         |
| configurations      | {                                                        |
|                     |   "router_id": "",                                       |
|                     |   "agent_mode": "legacy",                                |
|                     |   "gateway_external_network_id": "",                     |
|                     |   "handle_internal_only_routers": true,                  |
|                     |   "routers": 1,                                          |
|                     |   "interfaces": 1,                                       |
|                     |   "floating_ips": 1,                                     |
|                     |   "interface_driver": "neutron.agent.linux.interface.OVSInterfaceDriver", |
|                     |   "log_agent_heartbeats": false,                         |
|                     |   "external_network_bridge": "br-ex",                    |
|                     |   "ex_gw_ports": 1                                       |
|                     | }                                                        |
| created_at          | 2019-12-03 17:41:57                                      |
| description         |                                                          |
| heartbeat_timestamp | 2021-01-18 21:14:21                                      |
| host                | controller                                               |
| id                  | 26f38d05-3553-46f6-8740-847f97741f89                     |
| started_at          | 2021-01-18 09:02:50                                      |
| topic               | l3_agent                                                 |
+---------------------+----------------------------------------------------------+
[root@controller ~]#
```

图 5.44 查看 L3 agent 服务的详细信息

2. 网络创建和管理

（1）使用 openstack 帮助命令查看网络服务可以执行的操作，以及命令的具体使用参数与方法。

V5-5 网络创建和管理

```
[root@controller ~]# openstack help | grep network
                [--os-network-api-version <network-api-version>]
  --os-network-api-version <network-api-version>
  network create    Create new network
  network delete    Delete network(s)
  network list      List networks
  network set       Set network properties
  network show      Show network details
[root@controller ~]#
[root@controller ~]# openstack help | grep subnet
  subnet delete     Delete subnet
  subnet list       List subnets
  subnet pool delete    Delete subnet pool
  subnet pool list      List subnet pools
  subnet pool show      Display subnet pool details
  subnet show       Show subnet details
[root@controller ~]#
[root@controller ~]# openstack help network create
usage: openstack network create [-h]
                                [-f {html,json,json,shell,table,value,yaml,yaml}]
                                [-c COLUMN] [--max-width <integer>]
                                [--noindent] [--prefix PREFIX]
                                [--share | --no-share] [--enable | --disable]
                                [--project <project>]
                                [--project-domain <project-domain>]
                                [--availability-zone-hint <availability-zone>]
                                <name>
[root@controller ~]#
```

（2）使用 neutron 帮助命令查看网络服务可以执行的操作，以及命令的具体使用参数与方法。

```
[root@controller ~]# neutron help |grep net
                        Defaults to env[OS_NETWORK_SERVICE_TYPE] or network.
  bgp-speaker-network-add       Add a network to the BGP speaker.
  bgp-speaker-network-remove    Remove a network from the BGP speaker.
  dhcp-agent-list-hosting-net   List DHCP agents hosting a network.
  dhcp-agent-network-add        Add a network to a DHCP agent.
  dhcp-agent-network-remove     Remove a network from a DHCP agent.
  gateway-device-create         Create a network gateway device.
  gateway-device-delete         Delete a given network gateway device.
  gateway-device-list           List network gateway devices for a given tenant.
  gateway-device-show           Show information for a given network gateway device.
  gateway-device-update         Update a network gateway device.
  net-create                    Create a network for a given tenant.
  net-delete                    Delete a given network.
  net-external-list             List external networks that belong to a given tenant.
```

```
    net-gateway-connect              Add an internal network interface to a router.
    net-gateway-create               Create a network gateway.
    net-gateway-delete               Delete a given network gateway.
    net-gateway-disconnect           Remove a network from a network gateway.
    net-gateway-list                 List network gateways for a given tenant.
    net-gateway-show                 Show information of a given network gateway.
    net-gateway-update               Update the name for a network gateway.
    net-ip-availability-list         List IP usage of networks
    net-ip-availability-show         Show IP usage of specific network
    net-list                         List networks that belong to a given tenant.
    net-list-on-dhcp-agent           List the networks on a DHCP agent.
    net-show                         Show information of a given network.
    net-update                       Update network's information.
    router-gateway-clear             Remove an external network gateway from a router.
    router-gateway-set               Set the external network gateway for a router.
    router-interface-add             Add an internal network interface to a router.
    router-interface-delete          Remove an internal network interface from a router.
    subnet-create                    Create a subnet for a given tenant.
    subnet-delete                    Delete a given subnet.
    subnet-list                      List subnets that belong to a given tenant.
    subnet-show                      Show information of a given subnet.
    subnet-update                    Update subnet's information.
    subnetpool-create                Create a subnetpool for a given tenant.
    subnetpool-delete                Delete a given subnetpool.
    subnetpool-list                  List subnetpools that belong to a given tenant.
    subnetpool-show                  Show information of a given subnetpool.
    subnetpool-update                Update subnetpool's information.
[root@controller ~]#
[root@controller ~]# neutron help | grep subnet
    subnet-create                    Create a subnet for a given tenant.
    subnet-delete                    Delete a given subnet.
    subnet-list                      List subnets that belong to a given tenant.
    subnet-show                      Show information of a given subnet.
    subnet-update                    Update subnet's information.
    subnetpool-create                Create a subnetpool for a given tenant.
    subnetpool-delete                Delete a given subnetpool.
    subnetpool-list                  List subnetpools that belong to a given tenant.
    subnetpool-show                  Show information of a given subnetpool.
    subnetpool-update                Update subnetpool's information.
[root@controller ~]#
[root@controller ~]# neutron help subnet-create
usage: neutron subnet-create [-h]
                             [-f {html,json,json,shell,table,value,yaml,yaml}]
                             [-c COLUMN] [--max-width <integer>] [--noindent]
                             [--prefix PREFIX] [--request-format {json}]
                             [--tenant-id TENANT_ID] [--name NAME]
                             [--gateway GATEWAY_IP | --no-gateway]
                             [--allocation-pool start=IP_ADDR,end=IP_ADDR]
```

```
                    [--host-route destination=CIDR,nexthop=IP_ADDR]
                    [--dns-nameserver DNS_NAMESERVER]
                    [--disable-dhcp] [--enable-dhcp]
                    [--ip-version {4,6}]
                    [--ipv6-ra-mode {dhcpv6-stateful,dhcpv6-stateless,slaac}]
                    [--ipv6-address-mode {dhcpv6-stateful,dhcpv6-stateless,slaac}]
                    [--subnetpool SUBNETPOOL]
                    [--use-default-subnetpool]
                    [--prefixlen PREFIX_LENGTH]
                    NETWORK [CIDR]
[root@controller ~]#
[root@controller ~]# neutron help net-create
usage: neutron net-create [-h]
                    [-f {html,json,json,shell,table,value,yaml,yaml}]
                    [-c COLUMN] [--max-width <integer>] [--noindent]
                    [--prefix PREFIX] [--request-format {json}]
                    [--tenant-id TENANT_ID] [--admin-state-down]
                    [--shared] [--provider:network_type <network_type>]
                    [--provider:physical_network <physical_network_name>]
                    [--provider:segmentation_id <segmentation_id>]
                    [--vlan-transparent {True,False}]
                    [--qos-policy QOS_POLICY]
                    [--availability-zone-hint AVAILABILITY_ZONE]
                    [--dns-domain DNS_DOMAIN]
                    NAME
[root@controller ~]#
```

（3）使用 openstack 命令管理网络服务。

使用 openstack 命令创建内部网络，命令格式如下。

openstack network create 网络名称

例如，创建内部网络 in-net-01，其命令的结果如图 5.45 所示。

```
[root@controller ~]# openstack network create in-net-01
+---------------------------+--------------------------------------+
| Field                     | Value                                |
+---------------------------+--------------------------------------+
| admin_state_up            | UP                                   |
| availability_zone_hints   |                                      |
| availability_zones        |                                      |
| created_at                | 2021-01-19T08:53:05                  |
| description               |                                      |
| headers                   |                                      |
| id                        | 3a19016b-b536-4269-9c4a-bb467b480e94 |
| ipv4_address_scope        | None                                 |
| ipv6_address_scope        | None                                 |
| mtu                       | 1458                                 |
| name                      | in-net-01                            |
| port_security_enabled     | True                                 |
| project_id                | f9ff39ba9daa4e5a8fee1fc50e2d2b34     |
| provider:network_type     | gre                                  |
| provider:physical_network | None                                 |
| provider:segmentation_id  | 80                                   |
| router_external           | Internal                             |
| shared                    | False                                |
| status                    | ACTIVE                               |
| subnets                   |                                      |
| tags                      | []                                   |
| updated_at                | 2021-01-19T08:53:05                  |
+---------------------------+--------------------------------------+
[root@controller ~]#
```

图 5.45　使用 openstack 命令创建内部网络 in-net-01

使用 openstack 命令删除网络，命令格式如下。
openstack network delete 网络名称或 ID
例如，删除内部网络 in-net-01，其命令的结果如图 5.46 所示。

```
[root@controller ~]# openstack network list
+--------------------------------------+---------+--------------------------------------+
| ID                                   | Name    | Subnets                              |
+--------------------------------------+---------+--------------------------------------+
| 3a19016b-b536-4269-9c4a-bb467b480e94 | in-net-01|                                     |
| ce4678aa-a2c7-4399-a215-6907d929eaa5 | in-net  | 83cb607b-8e56-4d8f-b7b5-d84c375d660a |
| afddc555-61cb-4cf8-9bd4-1316079e6355 | out-net | 4207b78f-f5ab-4295-af9b-ae2a1ebe8032 |
+--------------------------------------+---------+--------------------------------------+
[root@controller ~]# openstack network delete in-net-01
[root@controller ~]# openstack network list
+--------------------------------------+---------+--------------------------------------+
| ID                                   | Name    | Subnets                              |
+--------------------------------------+---------+--------------------------------------+
| ce4678aa-a2c7-4399-a215-6907d929eaa5 | in-net  | 83cb607b-8e56-4d8f-b7b5-d84c375d660a |
| afddc555-61cb-4cf8-9bd4-1316079e6355 | out-net | 4207b78f-f5ab-4295-af9b-ae2a1ebe8032 |
+--------------------------------------+---------+--------------------------------------+
[root@controller ~]#
```

图 5.46　使用 openstack 命令删除网络 in-net-01

（4）使用 neutron 命令管理网络服务。

使用 neutron 命令创建内部网络，命令格式如下。
#neutron　net-create　网络名称
例如，创建内部网络 in-net-10，其命令的结果如图 5.47 所示。

```
[root@controller ~]# neutron net-create in-net-10
Created a new network:
+---------------------------+--------------------------------------+
| Field                     | Value                                |
+---------------------------+--------------------------------------+
| admin_state_up            | True                                 |
| availability_zone_hints   |                                      |
| availability_zones        |                                      |
| created_at                | 2021-01-19T09:42:37                  |
| description               |                                      |
| id                        | 38b67083-db83-4afb-9cae-dcbb2a31497e |
| ipv4_address_scope        |                                      |
| ipv6_address_scope        |                                      |
| mtu                       | 1458                                 |
| name                      | in-net-10                            |
| port_security_enabled     | True                                 |
| provider:network_type     | gre                                  |
| provider:physical_network |                                      |
| provider:segmentation_id  | 90                                   |
| router:external           | False                                |
| shared                    | False                                |
| status                    | ACTIVE                               |
| subnets                   |                                      |
| tags                      |                                      |
| tenant_id                 | f9ff39ba9daa4e5a8fee1fc50e2d2b34     |
| updated_at                | 2021-01-19T09:42:37                  |
+---------------------------+--------------------------------------+
[root@controller ~]#
[root@controller ~]# neutron net-list
+--------------------------------------+-----------+------------------------------------------------------+
| id                                   | name      | subnets                                              |
+--------------------------------------+-----------+------------------------------------------------------+
| ce4678aa-a2c7-4399-a215-6907d929eaa5 | in-net    | 83cb607b-8e56-4d8f-b7b5-d84c375d660a 192.168.100.0/24|
| 38b67083-db83-4afb-9cae-dcbb2a31497e | in-net-10 |                                                      |
| afddc555-61cb-4cf8-9bd4-1316079e6355 | out-net   | 4207b78f-f5ab-4295-af9b-ae2a1ebe8032 202.199.184.0/24|
+--------------------------------------+-----------+------------------------------------------------------+
[root@controller ~]#
```

图 5.47　使用 neutron 命令创建内部网络 in-net-10

使用 neutron 命令创建内部网络子网。例如，为内部网络"in-net-10"创建子网，子网名称为"sub_in-net-10"，网关地址为 192.168.10.254，子网起始地址为 192.168.10.100，结束地址为 192.168.10.200，分配网段为 192.168.10.0/24，执行命令如下。

#neutron　subnet-create in-net-10 --name　sub_in-net-10　--gateway 192.168.10.254 --allocation-pool start=192.168.10.100,end=192.168.10.200 192.168.10.0/24

其命令的结果如图 5.48 所示。

图 5.48　使用 neutron 命令创建内部网络子网 sub_in-net-10

使用 neutron 命令创建外部网络及其子网。例如，为外部网络"out-net-20"创建子网，子网名称为"sub_out-net-20"，网关地址为 202.199.20.254，子网起始地址为 202.199.20.100，结束地址为 202.199.20.200，分配网段为 202.199.20.0/24，执行命令如下。

　　# neutron　net-create out-net-20 --shared --router:external=True
　　#neutron　subnet-create out-net-20 --name　sub_out-net-20　--allocation-pool start=202.199.20.100,end=202.199.20.200　--disable-dhcp --gateway 202.199.20.254 202.199.20.0/24

其命令的结果如图 5.49 所示。

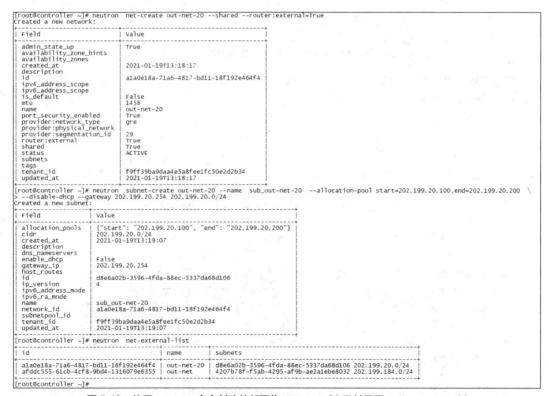

图 5.49　使用 neutron 命令创建外部网络 out-net-20 及其子网 sub_out-net-20

3. 路由创建和管理

（1）使用 openstack 命令管理路由，执行帮助命令如下。

```
[root@controller ~]# openstack help | grep router
  router create     Create a new router
  router delete     Delete router(s)
  router list       List routers
  router set        Set router properties
  router show       Display router details
[root@controller ~]#
```

使用 openstack 命令创建路由，命令格式如下。

```
#openstack  router  create  路由名称
```

例如，创建路由 router-10，其命令的结果如图 5.50 所示。

图 5.50　使用 openstack 命令创建路由 router-10

使用 openstack 命令删除路由，命令格式如下。

```
#openstack  router  delete  路由名称
```

其命令的结果如图 5.51 所示。

```
[root@controller ~]# openstack router delete router-10
[root@controller ~]# openstack router list
+--------------------------------------+-----------+--------+-------+-------------+-------+----------------------------------+
| ID                                   | Name      | Status | State | Distributed | HA    | Project                          |
+--------------------------------------+-----------+--------+-------+-------------+-------+----------------------------------+
| 0bd966c5-0d37-4f99-9d19-286e72e8c11e | router-01 | ACTIVE | UP    | False       | False | f9ff39ba9daa4e5a8fee1fc50e2d2b34 |
+--------------------------------------+-----------+--------+-------+-------------+-------+----------------------------------+
[root@controller ~]#
```

图 5.51　使用 openstack 命令删除路由 router-10

（2）使用 neutron 命令管理路由以及网关接口。

使用 neutron 命令创建路由以及网关接口，执行帮助命令如下。

```
[root@controller ~]# neutron help | grep router
  l3-agent-list-hosting-router      List L3 agents hosting a router.
  l3-agent-router-add               Add a router to a L3 agent.
  l3-agent-router-remove            Remove a router from a L3 agent.
  net-gateway-connect               Add an internal network interface to a router.
  router-create                     Create a router for a given tenant.
  router-delete                     Delete a given router.
  router-gateway-clear              Remove an external network gateway from a router.
  router-gateway-set                Set the external network gateway for a router.
  router-interface-add              Add an internal network interface to a router.
```

命令	说明
router-interface-delete	Remove an internal network interface from a router.
router-list	List routers that belong to a given tenant.
router-list-on-l3-agent	List the routers on a L3 agent.
router-port-list	List ports that belong to a given tenant, with specified router.
router-show	Show information of a given router.
router-update	Update router's information.

[root@controller ~]#

使用 neutron 命令创建路由，命令格式如下。

#neutron　router-create　路由名称

例如，创建路由 router-10，其命令的结果如图 5.52 所示。

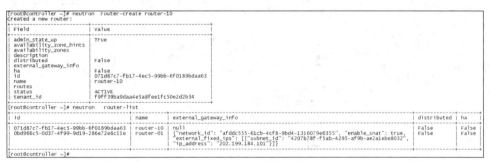

图 5.52　使用 neutron 命令创建路由 router-10

创建路由网关接口，命令格式如下。

neutron　router-interface-add　路由名称　内网子网名称或 ID
neutron router-list
neutron router-gateway-set　路由名称　外部网络名称或 ID
neutron router-list

例如，创建路由 router-10 的网关接口，其命令的结果如图 5.53 所示。

图 5.53　使用 neutron 命令创建路由 router-10 的网关接口

在 Dashboard 界面中，单击"项目"主节点，再单击"网络"→"网络拓扑"子节点，弹出网络拓扑界面，可以查看当前网络拓扑结构，如图 5.54 所示。

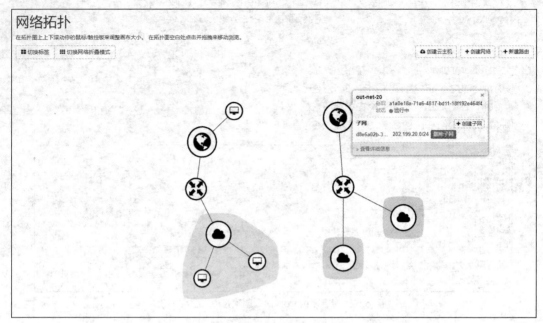

图 5.54　查看当前网络拓扑结构

课后习题

1. 选择题

（1）【多选】OpenStack 网络服务 Neutron 支持的网络拓扑类型有（　　）。
　　A. Local　　　　B. Flat　　　　　C. VLAN　　　　　D. GRE
（2）【多选】一个简化的 Neutron 网络结构，通常包括（　　）。
　　A. 一个外部网络（External Network）　B. 一个内部网络（Internal Network）
　　C. 一个路由器（Router）　　　　　　　D. 一个防火墙（Firewall）
（3）【多选】查看网络列表信息时，可使用的命令是（　　）。
　　A. openstack　network　list　　　　B. openstack　network-list
　　C. neutron　　network-list　　　　　D. neutron　　net-list
（4）【多选】查看某个网络的详细信息时，可使用的命令是（　　）。
　　A. openstack　network　show　网络名称或 ID
　　B. openstack　network-show　网络名称或 ID
　　C. neutron　net-show　网络名称或 ID
　　D. neutron　network-show　网络名称或 ID
（5）【多选】查看子网列表信息时，可使用的命令是（　　）。
　　A. openstack　subnet　list　　　　B. openstack　subnet　show
　　C. neutron　subnet-list　　　　　　D. neutron　subnet-show

(6)【多选】查看路由详细信息时,可使用的命令是（   ）。
  A. openstack router list
  B. neutron router-list
  C. openstack router show   路由名称或 ID
  D. neutron router- show   路由名称或 ID
(7)【多选】创建内部网络时,可使用的命令是（   ）。
  A. openstack network create 网络名称　B. openstack net create 网络名称
  C. neutron network-create 网络名称　D. neutron net-create 网络名称
(8)【多选】创建路由时,可使用的命令是（   ）。
  A. openstack router create 路由名称
  B. openstack router-create 路由名称
  C. neutron router create 路由名称
  D. neutron router-create 路由名称

2. 简答题

(1)简述 Linux 虚拟网桥的功能。
(2)简述开放虚拟交换机的功能。
(3)简述 Neutron 的网络结构。
(4)简述 Neutron 管理的网络资源。
(5)简述 Neutron 网络的拓扑类型。
(6)简述 Neutron 基本架构。
(7)简述 Neutron 的物理节点部署关系。
(8)简述 Neutron 的主要插件、代理与服务。

项目6
OpenStack计算服务

06

【学习目标】

- 掌握OpenStack计算服务基础。
- 掌握Nova部署架构。
- 掌握Nova的元数据工作机制。
- 掌握虚拟机实例管理。
- 掌握基于Web界面计算服务管理配置方法。
- 掌握基于命令行计算服务管理配置方法。

6.1 项目陈述

计算服务是 OpenStack 最核心的服务之一，负责维护和管理云环境的计算资源。计算服务是云计算的结构控制器，它是 IaaS 系统的主要部分，其主要模块由 Python 实现。计算服务在 OpenStack 中的项目代号为 Nova。Nova 可以说是 OpenStack 中最核心的组件，而 OpenStack 的其他组件，归根结底都是为 Nova 组件服务的。Nova 组件如此重要，注定它是 OpenStack 中最为复杂的组件。Nova 服务由多个子服务构成，这些子服务通过远程过程调用（Remote Procedure Call，RPC）实现通信。OpenStack 作为 IaaS 的云操作系统，通过 Nova 实现虚拟机生命周期管理。OpenStack 计算服务需要与其他服务进行交互，如身份服务用于认证、镜像服务提供磁盘和服务器镜像、Dashboard 提供用户与管理员接口。使用 OpenStack 管理虚拟机的方法已经非常成熟，通过 Nova 可以快速自动化地创建虚拟机。

6.2 必备知识

6.2.1 OpenStack 计算服务基础

OpenStack 计算服务是 IaaS 系统的重要组成部分，OpenStack 的其他组件依托 Nova，与 Nova 协同工作，组成了整个 OpenStack 云计算管理平台。OpenStack 使用它来托管和管理云计算系统。

1. 什么是 Nova

Nova 是 OpenStack 中的计算服务项目，计算实例（虚拟服务器）生命周期的所有活动都由

Nova 管理。Nova 支持创建虚拟机和裸金属服务器，并且支持系统容器。作为一套在现有 Linux 服务器上运行的守护进程，Nova 提供计算服务，但它自身并没有提供任何虚拟化能力，而是使用不同的虚拟化驱动来与底层支持的虚拟机管理器进行交互。Nova 需要下列 OpenStack 其他服务的支持。

（1）Keystone：这项服务为所有的 OpenStack 服务提供身份管理和认证。

（2）Glance：这项服务提供计算用的镜像库，所有的计算实例都从 Glance 镜像启动。

（3）Neutron：这项服务负责配置管理计算实例启动时的虚拟或物理网络连接。

（4）Cinder 和 Swift：这两项服务分别为计算实例提供块存储和对象存储支持。Nova 也能与其他服务集成，如加密磁盘和裸金属计算实例等。

V6-1 什么是 Nova

2. Nova 的系统架构

Nova 由多个提供不同功能的独立组件构成，对外通过 REST API 通信，对内通过 RPC 通信，使用一个中心数据库存储数据。其中，每个组件都可以部署一个或多个来实现横向扩展。Nova 的系统架构如图 6.1 所示。

V6-2 Nova 的系统架构

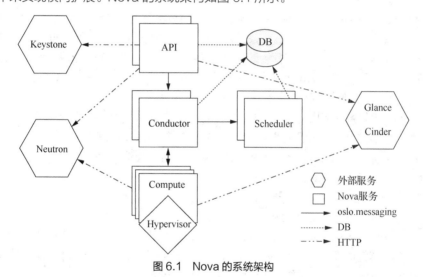

图 6.1 Nova 的系统架构

（1）API

API 是整个 Nova 组件的入口，用于接收和处理客户端发送的 HTTP 请求或 HTTP 与其他组件通信的 Nova 组件的信息等。

（2）Conductor

Conductor 是 OpenStack 中的一个 RPC 服务，主要提供对数据库的查询和权限分配操作，处理需要协调的请求（构建虚拟机或调整虚拟机大小）或是处理对象转换。

（3）Scheduler

Scheduler 是用于决定哪台主机承载计算实例的 Nova 调度器，可完成 Nova 的核心调度，包括虚拟机硬件资源调度、节点调度等。同时，Nova 的 Scheduler 决定了虚拟机创建的具体位置。

（4）Compute

Compute 是 Nova 的核心子组件，通过与 Nova 的 Client 进行交互，实现虚拟机的管理功能。它负责在计算节点上对虚拟机实例进行一系列操作，包括迁移、安全组策略和快照管理等。

（5）DB

DB 是用于数据存储的 SQL 数据库。

消息队列是 Nova 服务组件之间传递信息的中心枢纽，通常使用基于高级消息队列协议（Advanced Message Queuing Protocol，AMQP）的 RabbitMQ 消息队列来实现。为避免消息阻塞而造成长时间等待响应，Nova 计算服务组件采用了异步调用的机制，当请求被接收后，响应即被触发，发送回执，而不关注该请求是否完成。Nova 提供了虚拟网络，使实例之间能够彼此访问，可以访问公共网络。目前，Nova 的网络模块 nova-network 已经过时，现在已经使用 Neutron 网络服务组件。

>
>
> **注意** oslo.messaging 是 OpenStack Icehouse 消息处理架构。在 Icehouse 中，RPC 消息队列相关处理从 openstack.common.rpc 慢慢地转移到 oslo.messaging 上。这个架构更合理，代码结构清晰，且弱耦合。

3. API 组件

API 是客户端访问 Nova 的 HTTP 接口，它由 nova-api 服务实现，nova-api 服务接收和响应来自最终用户的计算 API 请求。作为 OpenStack 对外服务的最主要的接口，nova-api 提供了一个集中的可以查询所有 API 的端点。它是整个 Nova 组件的"门户"，所有对 Nova 的请求都首先由 nova-api 处理，API 提供 REST 标准调用服务，便于与第三方系统集成。可以通过运行多个 API 服务实例轻松实现 API 的高可用性。例如，运行多个 nova-api 进程，除了提供 OpenStack 自己的 API，nova-api 服务还支持 Amazon EC2 API。

最终用户不会直接发送 RESTful API 请求，而是通过 OpenStack 命令行、Dashboard 和其他需要与 Nova 交换的组件来使用这些 API。只要是与虚拟机生命周期相关的操作，nova-api 都可以响应。nova-api 对接收到的 HTTP API 请求会做出以下处理。

（1）检查客户端传入的参数是否合法有效。

（2）调用 Nova 其他服务的处理客户端 HTTP 请求。

（3）格式化 Nova 其他子服务返回的结果并返回给客户端。

nova-api 是外部访问并使用 Nova 提供的各种服务的唯一途径，也是客户端和 Nova 之间的中间层。它将客户端的请求传送给 Nova，待 Nova 处理请求之后再将处理结果返回给客户端。由于这种特殊地位，nova-api 被要求保持调度稳定，目前已经比较成熟和完备。

4. Conductor 组件

Conductor 组件由 nova-conductor 模块实现，旨在为数据库的访问提供一层安全保障。nova-conductor 是 OpenStack 中的一个 RPC 服务，主要提供数据库查询功能，Scheduler 组件只能读取数据库的内容，API 通过策略限制数据库的访问，两者都可以直接访问数据库，更加规范的方法是通过 Conductor 组件来对数据库进行操作。nova-conductor 作为 nova-compute 服务与数据库之间交互的中介，避免了直接访问由 nova-compute 服务创建的云数据库。nova-conductor 可以水平扩展，但是不要将它部署在运行 nova-compute 服务的节点上。

nova-compute 需要获取和更新数据库中虚拟机实例的信息。早期版本的 nova-compute 是直接访问数据库的，这可能带来安全和性能问题。从安全方面来说，如果一个计算节点被攻陷，数据库就会直接暴露出来；从性能方面来说，nova-compute 对数据库的访问是单线程、阻塞式的，而数据库处理是串行而不是并行的，这就会造成一个瓶颈问题。

使用 nova-conductor 可以解决这些问题。将 nova-compute 访问数据库的全部操作都放到 nova-conductor 中，nova-conductor 作为对此数据库操作的一个代理，而且 nova-conductor 是部署在控制节点上的。这样避免了 nova-compute 直接访问数据库，增加了系统的安全性。

使用 nova-conductor 也有助于提高数据库的访问性能。nova-compute 可以创建多个线程并使用 RPC 访问 nova-conductor。不过，通过 RPC 访问 nova-conductor 会受网络延迟的影响，且 nova-conductor 访问数据库是阻塞式的。

nova-conductor 将 nova-compute 与数据库分离之后提高了 Nova 的伸缩性。nova-compute 与 nova-conductor 是通过消息间接交互的。这种松散的架构允许配置多个 nova-conductor 实例。在一个大的 OpenStack 部署环境中，管理员可以通过增加 nova-conductor 的数量来应对日益增长的计算节点对数据库的访问。另外，nova-conductor 升级方便，在保持 Conductor API 兼容的前提下，数据库模式无须升级 nova-compute。

5. Scheduler 组件

Scheduler 可译为调度器，由 nova-scheduler 服务实现，旨在解决"如何选择在某个计算节点上启动实例"的问题。它应用多种规则，考虑内存使用率、CPU 负载率、CPU 构架等多种因素，根据一定的算法，确定虚拟机实例能够运行在哪一台计算服务器上。nova-scheduler 服务会从队列中接收一个虚拟机实例的请求，通过读取数据库的内容，从可用资源池中选择最合适的计算节点来创建新的虚拟机实例。

创建虚拟机实例时，用户会提出资源需求，如 CPU、内存、磁盘容量等。OpenStack 将这些需求定义在实例类型中，用户只需指定使用哪种实例类型就可以了。nova-scheduler 会按照实例类型选择合适的计算节点。

在/etc/nova/nova.conf 配置文件中，可通过 scheduler_dirver、scheduler_default_filters 和 scheduler_available_filters 这 3 个参数来配置 nova-scheduler。这里主要介绍 nova-scheduler 的调度机制和实现方法。

（1）Nova 调度类型

Nova 支持多种调度方式来选择运行虚拟机的主机节点，目前有以下 3 种调度器。

① 随机调度器：从所有 nova-compute 服务正常运行的节点中随机选择。

② 过滤调度器：根据指定的过滤条件以及权重选择最佳的计算节点。

③ 缓存调度器：可以看作随机调度器的一种特殊类型，在随机调度的基础上将主机资源信息缓存在本地内存中，然后通过后台的定时任务定时地从数据库中获取最新的主机资源信息。

调度器需要在/etc/nova/nova.conf 文件中通过 scheduler_driver 选项指定。为了便于扩展，Nova 将调度器必须要实现的一个接口提取出来，称为 nova.scheduler.driver.Scheduler，只要继承类 SchedulerDriver 并实现其中的接口，就可以实现自己的调度器。默认使用的是过滤调度器。

Nova 可使用第三方调度器，配置 scheduler_driver 即可，注意不同的调度器不能共存。

（2）过滤调度器调度过程

过滤调度器的调度过程分为两个阶段。

① 通过指定的过滤器选择满足条件的计算节点（运行 nova-compute 的主机），例如，内存使用率小于 50%，可以使用多个过滤器依次进行过滤。

② 对过滤之后的主机列表进行权重计算（Weighting）并排序，选择最优的（权重值最大）计算节点来创建虚拟机实例。

在一台高性能主机上创建一台功能简单的普通虚拟机的代价是较大的，OpenStack 对权限值的计算需要一个或多个（Weight 值，代价函数）的组合，然后对每一个经过滤的主机调用代价函数进行计算，将得到的值与 Weight 值相乘，得到最终的权值。OpenStack 将在权值最大的主机上创建虚拟机实例。

过滤调度器调度过程如图 6.2 所示。刚开始有 6 台可用的计算节点主机，通过多个过滤器层层过滤，将主机 2、主机 3 和主机 5 排除。剩下的 3 台主机再通过计算权重与排序，按优先级从高到低依次为主机 4、主机 1 和主机 6。主机 4 权重值最高，最终入选。

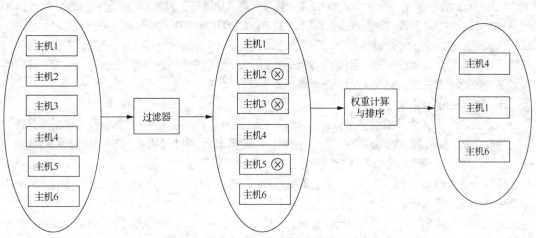

图 6.2　过滤调度器调度过程

（3）过滤器

当过滤调度器需要执行调度操作时，会让过滤器对计算节点进行判断，返回 True 或 False。/etc/nova/nova.conf 配置文件中的 scheduler_available_filters 选项用于配置可用的过滤器，默认所有 Nova 自带的过滤器都可以用于过滤操作。

另外，还有一个选项 scheduler_default_filters，用于指定 nova-scheduler 服务真正使用的过滤器，过滤调度器将按照列表中的顺序依次过滤，默认顺序为再审过滤器（RetryFilter）、可用区域过滤器（AvailabilityZoneFilter）、内存过滤器（RamFilter）、磁盘过滤器（DiskFilter）、核心过滤器（CoreFilter）、计算过滤器（ComputeFilter）、计算能力过滤器（ComputeAbilityFilter）、镜像属性过滤器（ImagePropertiesFilter）、服务器组反亲和性过滤器（ServerGroupAntiAffinityFilter）和服务器组亲和性过滤器（ServerGroupAffinityFilter）。

① 再审过滤器。再审过滤器的作用是过滤之前已经调度过的节点，例如，主机 1、主机 2 和主机 3 都通过了过滤，最终主机 1 因为权重值最大被选中执行操作，但由于某种原因，操作在主机 1 上执行失败了。默认情况下，nova-scheduler 会重新执行过滤操作（重复次数由 scheduler_max_attempts 选项指定，默认值是 3）。此时，RetryFilter 就会将主机 1 直接排除，避免操作再次失败。RetryFilter 通常作为第一个过滤器使用。

② 可用区域过滤器。可用区域过滤器为提高容灾性并提供隔离服务，可以将计算节点划分到不同的可用区域中。OpenStack 默认有一个命令为 Nova 的可用区域，所有的计算节点初始都是放在 Nova 中的。用户可以根据需要创建自己的可用区域。创建实例时，需要指定将实例部署在哪个可用区域中。nova-scheduler 执行过滤操作时，会使用可用区域过滤器将不属于指定可用区域的计算节点过滤掉。

③ 内存过滤器。内存过滤器根据可用内存来调度虚拟机创建，将不能满足实例类型内存需求的计算节点过滤掉。值得注意的是，为提高系统的资源使用率，OpenStack 在计算节点的可用内存时允许超过实际内存大小，超过的程序是通过 nova.conf 配置文件中的 ram_allocation_ratio 参数来控制的，默认值为 1.5。按照这个比例，假如计算节点的内存为 16GB，OpenStack 则会认为它有 24GB（16×1.5）的内存。

④ 磁盘过滤器。磁盘过滤器根据可用磁盘空间来调度虚拟机创建，将不能满足实例类型磁盘需求的计算节点过滤掉。对磁盘同样允许超量，可通过 nova.conf 中的 disk_allocation_ratio 参数控制，默认值为 1.0。

⑤ 核心过滤器。核心过滤器根据可用 CPU 核心来调度虚拟机创建，将不能满足实例类型 vCPU 需求的计算节点过滤掉。对 vCPU 同样允许超量，可通过 nova.conf 中的 cpu_allocation_ratio 参数控制，默认值为 16.0。

按照这个超量比例，nova-scheduler 在调度时会认为一个拥有 10 个 vCPU 的计算节点有 160 个 vCPU。不过 nova-scheduler 默认使用的过滤器并不包含核心过滤器。如果要使用，可以将核心过滤器添加到 nova.conf 的 scheduler_default_filters 配置选项中。

⑥ 计算过滤器。计算过滤器保证只有 nova-compute 服务正常工作的计算节点才能够被 nova-scheduler 调度，它显然是必选的过滤器。

⑦ 计算能力过滤器。这个过滤器可根据计算节点的特性来过滤。例如，x86_64 和 ARM 架构的不同节点，要将实例指定到部署 x86_64 架构的节点上，就可以利用该过滤器。

⑧ 镜像属性过滤器。镜像属性过滤器根据所选镜像的属性来筛选匹配的计算节点。与实例类似，镜像也有元数据，用于指定其属性。例如，希望某个镜像只能运行在 KVM 的 Hypervisor 上，可以通过 Hypervisor Type 属性来指定。如果没有设置镜像元数据，则镜像属性过滤器不会起作用，所有节点都会通过筛选。

⑨ 服务器组反亲和性过滤器。这个过滤器要求尽量将实例分散部署到不同的节点上。例如，有 4 个实例 s1、s2、s3 和 s4，4 个计算节点 A、B、C 和 D，为保证分散部署，会将 4 个实例 s1、s2、s3 和 s4 分别部署到不同计算节点 A、B、C 和 D 上。

⑩ 服务器组亲和性过滤器。与服务器组反亲和性过滤器的作用相反，此过滤器尽量将实例部署到同一个计算节点上。

6. Compute 组件

调度服务只负责分配任务，真正执行任务的是工作服务 Worker。在 Nova 中，这个 Worker 就是 Compute 组件，由 nova-compute 服务实现。这种职能划分使得 OpenStack 非常容易扩展。一方面，当计算资源不够，无法创建实例时，可以增加计算节点；另一方面，当客户的请求量太大，调度不过来时，可以增加调度器部署。

nova-compute 在计算节点上运行，负责管理节点上的实例，通常一台主机运行一个 nova-compute 服务，一个实例部署在哪台可用的主机上取决于调度算法。OpenStack 对实例的操作，最后都是交给 nova-compute 来完成的。nova-compute 的功能可以分为两类，一类是定时向 OpenStack 报告计算节点的状态，另一类是实现实例生命周期的管理。

（1）通过驱动架构支持多种 Hypervisor

创建虚拟机实例最终需要与 Hypervisor 打交道。Hypervisor 是计算节点上运行的虚拟化管理程序，也是虚拟机管理最底层的程序。不同虚拟化技术提供自己的 Hypervisor，常用的 Hypervisor 有 KVM、Xen、VMware 等。nova-compute 与 Hypervisor 一起实现 OpenStack 对实例生命周期的管理。它通过 Hypervisor 的 API（虚拟化层 API）来实现创建和销毁虚拟机实例的 Worker

守护进程，例如，XenServer/XCP 的 Xen API、KVM 或 QEMU 的 Libvirt 和 VMware 的 VMware API。这个处理过程相当复杂。基本上，该守护进程接收来自队列的动作请求，并执行一系列系统命令，如启动一个 KVM 实例并在数据库中更新它的状态。

面对多种 Hypervisor，nova-compute 为这些 Hypervisor 定义统一的接口，Hypervisor 只需要实现这些接口，就可以以 Driver 的形式即插即用到 OpenStack 系统中。nova-compute 的驱动架构如图 6.3 所示。

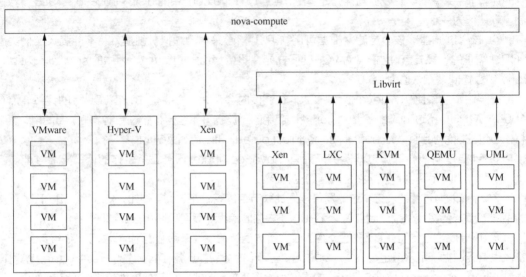

图 6.3 nova-compute 的驱动架构

一个计算节点上只能运行一种 Hypervisor，在该节点 nova-compute 的配置文件/etc/nova/nova.conf 中配置对应的 compute_driver 参数即可。例如，通常使用 KVM，配置 Libvirt 的驱动即可。

（2）定期向 OpenStack 报告计算节点的状态

OpenStack 通过 nova-compute 的定期报告获知每个计算节点的信息。每隔一段时间，nova-compute 就会报告当前计算节点的资源使用情况和 nova-compute 的服务状态。nova-compute 是通过 Hypervisor 的驱动获取这些信息的。例如，如果使用 Hypervisor 的是 KVM，则会使用 Libvirt 驱动，由 Libvirt 驱动调用相关的 API 获得资源信息。

（3）实现虚拟机实例生命周期的管理

OpenStack 对虚拟机实例最主要的操作都是通过 nova-compute 实现的，包括实例的创建、关闭、重启、挂起、恢复、中止、调整大小、迁移、快照等。

这里以实例创建为例来说明 nova-compute 的实现过程。当 nova-scheduler 选定部署实例的计算节点后，会通过消息中间件 RabbitMQ 向所选的计算节点发出创建实例的命令。计算节点上运行的 nova-compute 收到消息后会执行实例创建操作，创建过程可以分为以下几个阶段。

① 为实例准备资源。nova-compute 会根据指定的实例类型依次为要创建的实例分配 vCPU、内存和磁盘空间。

② 创建实例的镜像文件。OpenStack 创建一个实例时会选择一个镜像，这个镜像由 Glance

管理。nova-compute 首先用 Glance 将指定的镜像下载到计算节点，然后以其作为支持文件创建实例的镜像文件。

③ 创建实例的可扩展标记语言（eXtensible Markup Language，XML）定义文件。

④ 创建虚拟网络并启动虚拟机。

7. 虚拟机实例化流程

下面以创建虚拟机为例说明虚拟机实例化流程。

（1）用户执行 Nova Client 提供的用于创建虚拟机的命令。

（2）nova-api 服务监听来自 Nova Client 的 HTTP 请求，并将这些请求转换为 AMQP 消息之后加入消息队列。

（3）通过消息队列调用 nova-conductor 服务。

（4）nova-conductor 服务从消息队列中接收虚拟机实例化请求消息后，进行一些准备工作，例如，汇总 HTTP 请求中所需要实例化的虚拟机参数。

（5）nova-conductor 服务通过消息队列告诉 nova-scheduler 服务去选择一个合适的计算节点来创建虚拟机，此时 nova-scheduler 会读取数据库的内容。

（6）nova-conductor 服务从 nova-scheduler 服务得到了合适的计算节点的信息后，通过消息队列来通知 nova-compute 服务实现虚拟机的创建。

从虚拟机实例化的过程可以看出，Nova 中最重要的服务之间的通信是由消息队列来实现的，这符合松耦合的实现方式。并不是所有的业务流程都像创建虚拟机那样需要所有的服务来配合，例如，删除虚拟机实例时，就不需要 nova-conductor 服务，API 通过消息队列通知 nova-compute 服务删除指定的虚拟机，nova-compute 服务再通过 nova-conductor 服务更新数据库，至此完成该流程。

6.2.2 Nova 部署架构

Nova 使用基于消息、无共享、松耦合、无状态的架构，因而其部署非常灵活。

1. Nova 物理部署

OpenStack 是一个无中心结构的分布式系统，其物理部署非常灵活，可以部署到多个节点上，以获得更好的性能和高可用性。当然，也可以将所有的服务都安装在一台物理机上作为一个 All-in-One 测试环境。Nova 只是 OpenStack 的一个子系统，由多个组件和服务组成，可将它们部署在计算节点和控制节点这两类节点上。在计算节点上安装 Hypervisor 以运行虚拟机，只安装 nova-compute 即可。其他 Nova 组件和服务等则一起部署在控制节点上，如 nova-api、nova-scheduler、nova-conductor 等，以及 RabbitMQ 和 SQL 数据库。

客户端使用计算实例并不是直接访问计算节点的，而是通过控制节点提供的 API 来访问的。如果一个控制节点同时作为一个计算节点，则需要在上面运行 nova-compute。

通过增加控制节点和计算节点可实现简单方便的系统扩容。Nova 是可以水平扩展的，可以将多个 nova-api、nova-conductor 部署在不同节点上以提高服务能力，也可以运行多个 nova-scheduler 来提高可靠性。

Nova 经典部署模式是一个控制节点对应多个计算节点，如图 6.4 所示。

Nova 负载均衡部署模式则通过部署多个控制节点实现，如图 6.5 所示，当多个节点运行 nova-api 时，要在前端做负载均衡。

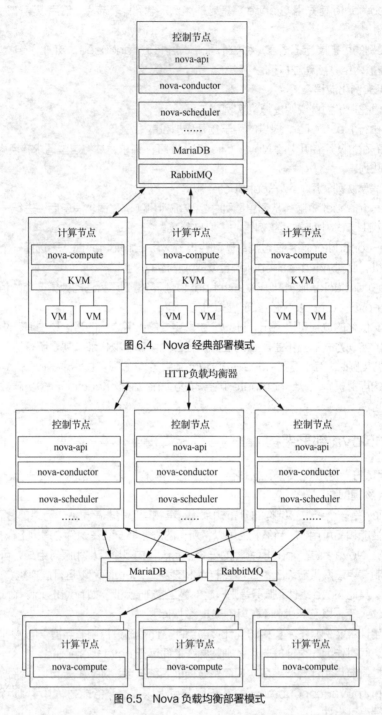

图 6.4 Nova 经典部署模式

图 6.5 Nova 负载均衡部署模式

当多个节点运行 nova-scheduler 或 nova-conductor 时,可由消息队列服务实现负载均衡。

2. Nova 的 Cell 架构

当 OpenStack 的 Nova 集群规模变大时,数据库和消息队列服务就会出现性能瓶颈问题。为提高水平扩展以及分布式、大规模部署能力,同时不增加数据库和消息中间件的复杂度,Nova 从

OpenStack 的 Grzzly 版本开始引入 Cell 的概念。

Cell 可译为单元。为支持更大规模的部署，OpenStack 将大的 Nova 集群分成小的单元，每个单元都有自己的消息队列和数据库，可以解决规模增加时引起的性能问题。Cell 不会像 Region 那样将各个集群独立运行。在 Cell 中，Keystone、Neutron、Cinder、Glance 等资源是可共享的。

早期的 Cell v1 版本的设想很好，但是局限于早期 Nova 架构，增加了一个 nova-cell 服务在各个单元之间传递消息，使架构更加复杂，没有被推广开来。OpenStack 从 Pike 版本开始弃用 Cell v1 版本。现在部署的都是 Cell v2 版本，它从 Newton 版本开始被引入，从 Ocata 版本开始变为必要组件，默认部署都会初始化一个单 Cell 的架构。

（1）Cell v2 的架构

Cell v2 的架构如图 6.6 所示，所有的 Cell 形成一个扁平架构，API 与 Cell 节点之间存在边界。API 节点只需要数据库，不需要消息队列。nova-api 依赖 nova-api 和 nova-cell0 两个数据库。API 节点上部署 nova-scheduler 服务，在调度的时候只需要在数据库中查出对应的 Cell 信息就能直接连接过去，从而出现一次调度就可以确定具体在哪个 Cell 的哪台机器上启动。在 Cell 节点中只需要安装 nova-compute 和 nova-conductor 服务，以及它依赖的消息队列和数据库。API 上的服务会直接连接 Cell 的消息队列和数据库，不需要像 nova-cell 这样的额外服务。Cell 下的计算节点只需要注册到所在的 Cell 节点下就可以了。

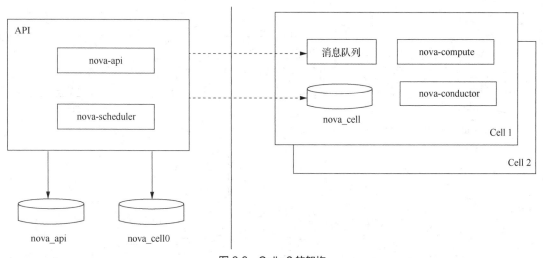

图 6.6　Cell v2 的架构

（2）API 节点上的数据库

根据 Cell v2 的设计，API 节点只使用两个数据库实例，即 nova_api 和 nova_cell0。nova_api 数据库中存放全局信息，这些全局信息是从 nova 库迁过来的，如实例模型（flavor）、实例组（instance groups）、配额（quota）等。

其中，Cell 的信息存放在 cell_mappings 表（存放 Cell 的数据库和消息的连接）中，用于与子 Cell 通信。主机的信息存放在 host_mappings 表中，用于 nova-scheduler 的调度，确认分配到的节点。实例的信息存放在 instance_mappings 表中，用于查询实例所在的 Cell，并连接 Cell 获取该实例的具体信息。

nova_cell0 数据库的模式与 Nova 的一样，当实例调度失败时，实例的信息不属于任何一个

Cell，因而存放在 nova_cell0 数据库中。

（3）Cell 的部署

基本的 Nova 系统包括以下组件。

① nova-api 服务：对外提供 REST API。

② nova-scheduler 服务：服务跟踪资源，决定实例放在哪个计算节点上。

③ API 数据库：主要用 nova-api 和 nova-scheduler（以下称为 API 层服务）跟踪实例的位置信息，以及正在构建但还未完成调度的实例的临时性位置。

④ nova-conductor 服务：卸载 API 层服务长期运行的任务，避免计算节点直接访问数据。

⑤ nova-compute 服务：管理虚拟机驱动和 Hypervisor 主机。

⑥ Cell 数据库：由 API、nova-conductor 和 nova-compute 服务使用，存放实例主要信息。

⑦ Cell0 数据库：与 Cell 数据库非常类似，但是仅存放那些调度失败的实例信息。

⑧ 消息队列：让服务之间通过 RPC 进行相互通信。

所有的部署都必须包括上述组件，小规模部署可能会使单一消息队列被各服务共享，单一数据库服务器承载 API 数据库、单个 Cell 数据库和必需的 Cell0 数据库，这通常被称为单 Cell 的部署，因为只有一个实际的 Cell，如图 6.7 所示。Cell0 数据库模拟一个正常的 Cell，但是没有计算节点，仅用于存储部署到实际计算节点失败的实例。所有服务配置通过同一消息总线进行相互通信，只有一个 Cell 数据库用于存储实例信息。

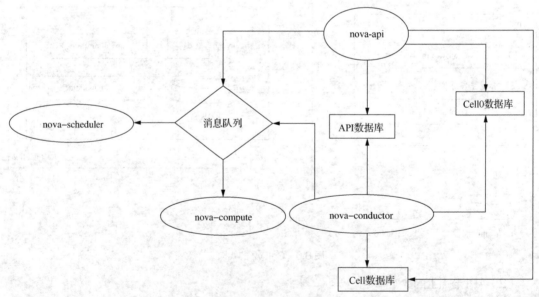

图 6.7　单 Cell 的部署

Nova 中的 Cell 架构的目的就是支持大规模部署，将许多计算节点划分到若干 Cell 中，每个 Cell 都有自己的数据库和消息队列。API 数据库只能是全局性的，有许多 Cell 数据库用于大量实例信息，每个 Cell 数据库承担整体部署中的实例的一部分。多 Cell 的部署如图 6.8 所示。首先，消息总线必须分割，Cell 数据库拥有同一总线。其次，必须为 API 层服务运行专用的 Conductor 服务，让它访问 API 数据库和专用的消息队列。我们将它称为超级引导器（Super Conductor），以便同每个 Cell 引导器节点明确区分。

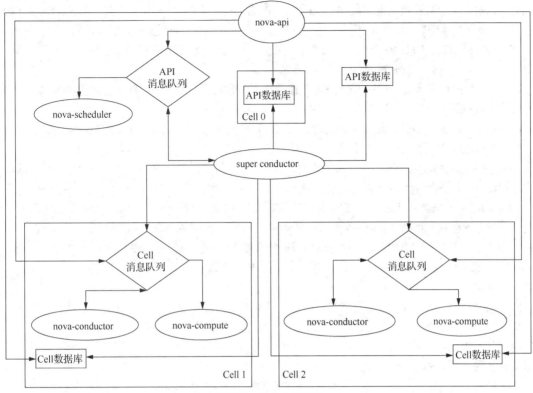

图 6.8 多 Cell 的部署

6.2.3 Nova 的元数据工作机制

IaaS 云计算管理平台中虚拟机启动时的自定义配置是必不可少的功能，OpenStack 通过一种称为元数据的机制来实现这一功能。虚拟机的元数据是一组与一台虚拟机相关联的键值对。Nova 支持为实例提供定制的元数据，目的是为启动的实例提供配置信息和设置参数。实例可通过 nova-api-metadata 服务或者配置驱动器（Config Driver）这两种途径获取元数据，而如何使用这些元数据由 cloud-init 负责。

1. 元数据及其注入

OpenStack 的实例是基于镜像部署的，镜像中包含了操作系统、最常用的软件以及最通用的配置。在实际应用中，创建实例时通常要对实例进行一些额外的自定义配置，如安装软件包、添加 SSH 密钥、配置主机名、设置磁盘容量、执行脚本等。如果可以将这些自定义配置都加到镜像中，则每次部署虚拟机实例都要制定定制化的镜像，不仅费时费力，而且大量、庞杂的镜像不便于管理，违背了镜像作为模板提供通用配置的初衷。当然，也可以在实例创建之后手动完成这些个性化配置，只是这不符合云服务自动化的基本要求。可行的方案是通过向虚拟机实例注入元数据信息，实例启动时获得自己的元数据，实例中的 cloud-init 工具根据元数据完成个性化配置工作。这样就不需要修改基础镜像了，在保证镜像稳定性的同时实现实例的自动化个性配置，从而不用单独为每一个虚拟机实例进行手动初始化操作。

OpenStack 将 cloud-init 定制虚拟机实例配置时获取的元数据信息分成两大类：元数据和用户数据。这两大类数据只代表了不同的信息类型，实质上都是提供配置信息的数据源，使用了相同的信息注入机制。元数据是结构化数据，以键值对形式注入实例，包括实例自身的一些常用属性，如主机名、网络配置信息、SSH 密钥等。用户数据是非结构化数据，通过文件或脚本的方式进行注入，支持多种文件格式，如 Shell 脚本、GZIP 压缩文件、cloud-init 配置文件等，主要包括一些命令、脚本等，例如，提供 Shell 脚本、设置 root 密码。

OpenStack 将元数据和用户数据的配置信息注入机制分为两种：一种是配置驱动器机制，另一种是元数据服务机制。

可以说，cloud-init 工具与配置驱动器机制或元数据服务机制一起实现了虚拟机实例的个性化定制。下面以 SSH 密钥注入为例说明元数据的实现过程。

（1）OpenStack 创建一个 SSH 密钥对，将其公钥（Public Key）存放在 OpenStack 数据库中，而将其私钥（Private Key）提供给用户，用户可以下载。

（2）创建一个实例时选择该 SSH 密钥对，完成实例创建之后，cloud-init 将其中的公钥写入实例，一般会保存到 .ssh/authorized_keys 目录中。

（3）用户可以用该 SSH 密钥对的私钥直接登录该实例。

2. 配置驱动器

（1）实现机制

OpenStack 将元数据信息写入实例的一个特殊的配置设备中，即配置驱动器中存储元数据，在实例启动时，自动挂载该设备，并由 cloud-init 读取其中的元数据信息，从而实现配置信息注入。任何可以挂载 ISO 9660 或者 VFAT 文件系统的客户操作系统，都可以使用配置驱动器，可以说配置驱动器是一种特殊的文件系统。

配置驱动器的具体实现根据 Hypervisor 的不同和配置会有所不同，不同的底层 Hypervisor 支持所挂载的设备类型也不尽相同。下面以常用的 Libvirt 为例进行说明。OpenStack 将元数据写入 Libvirt 的虚拟磁盘文件中，并指示 Libvirt 将其虚拟为 CD-ROM 设备。在启动实例时，客户操作系统中的 cloud-init 会挂载并读取该设备，并根据所读取的内容对实例进行配置。其中的 user_data 文件就是在创建虚拟机实例时指定需要执行的脚本文件。

（2）计算主机和镜像的要求

使用配置驱动器对计算机主机和镜像有一定的要求。

① 计算机主机必须符合以下要求。

Hypervisor 可以是 Libvirt、XenServer、Hyper-V 或 VMware，以及裸金属服务。与 Libvirt、XenServer、Hyper-V 或 VMware 一起使用配置驱动器时，必须先在计算机主机上安装 genisoimage 包，否则实例将不会被正确引导。与裸金属服务一起使用配置驱动器时，裸金属服务已经配置好。

② 镜像必须符合以下要求。

尽可能采用最新版本的 cloud-init 包制作镜像。如果一个镜像没有安装 cloud-init 包，那么必须定制镜像运动脚本，实现在实例启动期间挂载配置磁盘、读取数据、解析数据并且根据数据内容执行相应动作。如果将 Xen 与一个配置驱动器一起使用，要使用 xenapi_disable_agent 配置参数来禁用代理服务。在客户操作系统中，存储元数据的设备需要是 ISO 9660 或者 VFAT 文件系统。

当创建访问配置驱动器中的数据的镜像，并且 OpenStack 目录下有多个目录时，要选择所支持的最新的 API 版本。

（3）启用和访问配置驱动器

要启用配置驱动器，需要将 --config-drive true 参数传入 openstack server create 命令。

（4）配置驱动器格式

配置驱动器默认是 ISO 9660 文件系统。可在/etc/nova/nova.conf 文件中明确定义，命令如下。

config_drive_format=iso9660

默认情况下不能将配置驱动器镜像作为一个 CD-ROM 来替换磁盘驱动器。添加 CD-ROM 时，可以在/etc/nova/nova.conf 文件中定义，命令如下。

config_drive_cdrom=true

配置驱动器支持两种格式：ISO 9660 和 VFAT。其默认是 ISO 9660，但这会导致实例无法在线迁移，必须设置成 config_drive_format=vfat 才能在线迁移，这一点需要注意。

（5）应用场合

如果实例无法通过 DHCP 正确获取网络信息，使用配置驱动器就非常必要。配置驱动器主要用于配置实例的网络信息，包括 IP 地址、子网掩码、网关等。例如，可以先通过配置驱动器来给一个实例传递 IP 地址配置，这样在配置这个实例的网络设置之前，就可以先加载和访问这个实例了。

3. 元数据服务

如果实例能够自动正确配置网络，则可以通过元数据服务的方式获取元数据信息。OpenStack 提供 RESTful 接口让虚拟机实例通过 REST API 来获取元数据，这主要是由 nova-api-metadata 组件实现的，同时需要 neutron-metadata-agent 和 neutron-ns-metadata-proxy 这两个组件的配合。元数据服务的基本架构如图 6.9 所示。

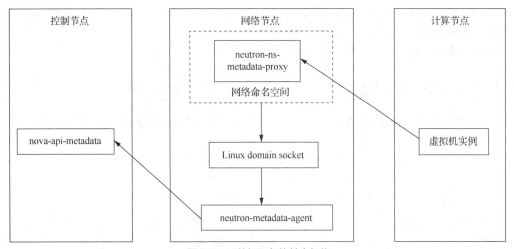

图 6.9　元数据服务的基本架构

这 3 个组件以及它们之间的关系说明如下。

（1）nova-api-metadata。运行在控制节点上，启动 RESTful 服务，负责处理实例发送的 REST API 请求。从请求 HTTP 头部中取出相应的信息，获取实例的 ID 继而从数据库中读取实例的元数据信息，最后将结果返回。作为元数据的提供者，nova-api-metadata 是 nova-api 的一个子服务，实例正是通过 nova-api-metadata 的 REST API 来获取元数据信息的。

（2）neutron-metadata-agent。运行在网络节点上，负责将接收到的获取元数据请求转发给 nova-api-metadata。neutron-metadata-agent 会获取实例和项目的 ID，并将其添加到请求的 HTTP 头部中，让 nova-api-metadata 根据这些信息获取元数据。

（3）neutron-ns-metadata-proxy。运行在网络节点上。如果由 DHCP 代理（neutron-dhcp-agent）创建，则它运行在 DHCP 代理所在的网络命名空间中；如果由 L3（neutron-l3-

agent）代理创建，则它运行在 Neutron 路由器所在的网络命名空间中。因为实例获取元数据的请求都是以路由器和 DHCP 服务器作为网络出口的，所以需要通过 neutron-ns-metadata-proxy 连通不同的网络命名空间，将请求在网络命名空间之间转发。neutron-ns-metadata-proxy 利用在 Linux domain socket 之上的 HTTP 技术，实现不同网络命名空间的 HTTP 请求转发。这样就在实例和 nova-api-metadata 之间建立起了通信。

nova-api-metadata 在控制节点上使用的是管理网络，由于网络不通，实例无法直接访问元数据服务，转而借助 neutron-metadata-agent 将请求转到 nova-api-metadata。而 neutron-metadata-agent 使用的也是管理网络，这样实例不能与 neutron-metadata-agent 通信，不过网络节点上有另外两个组件——DHCP 代理和 L3 代理，它们与实例可以位于同一 OpenStack 网络中，而 DHCP 服务器和 Neutron 路由器都在各自独立的网络命名空间中，于是可以通过 neutron-ns-metadata-proxy 解决连通网络命名空间和通信的问题。

下面总结一下虚拟机实例获取元数据的流程。

① 实例通过项目网络将元数据请求发送到 neutron-ns-metadata-proxy，此时会在请求中添加 router-id 和 network-id。

② neutron-ns-metadata-proxy 通过 Linux domain socket 将请求发送给 neutron-metadata-agent。此时根据请求中的 router-id、network-id 和 IP 地址获取端口信息，从而获得 instance-id 和 project-id 并加入请求。

③ neutron-metadata-agent 通过内部管理网络将请求转发给 nova-api-metadata。此时利用 instance-id 和 project-id 获取实例的元数据。

④ 将获取的元数据原路返回给发出请求的实例。

6.2.4 虚拟机实例管理

对于最终用户来说，可以使用以下工具，或者直接使用 API 通过 Nova 创建和管理服务器（虚拟机实例）。

（1）Horizon（Dashboard）。OpenStack 项目的 Web 图形界面。

（2）OpenStack 客户端。OpenStack 项目的命令行。

（3）Nova 客户端。对于 Nova 的一些高级特性和管理命令而言，需要使用该工具。

OpenStack 目前虽然支持 Nova 命令行，但是推荐使用 OpenStack 客户端。OpenStack 客户端不仅包括 Nova 命令，还提供 OpenStack 项目的绝大多数命令。

1. 部署虚拟机实例的前提

在创建实例之前需要做一些准备，确定以下基本要素。

（1）源。源是用来创建实例的模板，可以使用一个镜像、一个实例的快照、一个卷或一个卷快照，也可以通过创建一个新卷来选择使用具有持久性的存储。

（2）实例类型。实例类型也就是实例规格，定义了实例可使用的 CPU、内存和存储容量等硬件资源。

（3）密钥对。密钥对允许用户使用 SSH 访问新创建的实例。大部分云镜像支持公钥认证而不是传统的密码认证。在创建实例时，必须将一个公共密钥添加到计算服务中。

（4）安全组。通过访问规则定义防火墙策略，可以是提供者网络和私有网络。

2. 创建虚拟机实例

普通用户一般会通过基于 Web 的 Dashboard 图形界面创建虚拟机实例。Dashboard 图形界面中有两种创建实例的入口,一种是从镜像创建实例,另一种是直接创建实例。

管理员或测试人员一般会基于命令行创建虚拟机实例,建议使用 OpenStack 命令来代替 Nova 命令,执行命令行操作首先需要加载用户凭据的环境变量。

创建一个实例时,必须至少指定一种实例类型、镜像名称、网络、安全组、密钥和实例名称。因此创建实例之前,可以执行以下命令查看这些要素。

```
openstack  flavor  list                  //列出可用的实例类型
openstack  image  list                   //列出可用的镜像
openstack  network  list                 //列出可用的网络
openstack  security  group  list         //列出可用的安全组
openstack  keypair  list                 //列出可用的密钥对
```

执行创建实例命令,如下。

```
openstack server create --flavor m1.tiny --image cirros --nic net-id=public  \
--security-group default  --key-name demo-key  cirros-10
```

由于没有启动卷,这个基于镜像创建的虚拟机实例本身将存储在临时磁盘中,一旦重启,实例运行时添加或修改的数据将会丢失。

实例创建命令 openstack server create 的详细语法如下。

```
positional arguments:
    <server-name>           New server name
optional arguments:
    -h, --help              show this help message and exit
    --image <image>         Create server from this image (name or ID)
    --volume <volume>       Create server from this volume (name or ID)
    --flavor <flavor>       Create server with this flavor (name or ID)
    --security-group <security-group-name>
                            Security group to assign to this server (name or ID)
                            (repeat for multiple groups)
    --key-name <key-name>
                            Keypair to inject into this server (optional
                            extension)
    --property <key=value>
                            Set a property on this server (repeat for multiple
                            values)
    --file <dest-filename=source-filename>
                            File to inject into image before boot (repeat for
                            multiple files)
    --user-data <user-data>
                            User data file to serve from the metadata server
    --availability-zone <zone-name>
                            Select an availability zone for the server
    --block-device-mapping <dev-name=mapping>
                            Map block devices; map is
                            <id>:<type>:<size(GB)>:<delete_on_terminate> (optional
                            extension)
    --nic
<net-id=net-uuid,v4-fixed-ip=ip-addr,v6-fixed-ip=ip-addr,port-id=port-uuid>
                            Create a NIC on the server. Specify option multiple
```

	times to create multiple NICs. Either net-id or port-id must be provided, but not both. net-id: attach NIC to network with this UUID, port-id: attach NIC to port with this UUID, v4-fixed-ip: IPv4 fixed address for NIC (optional), v6-fixed-ip: IPv6 fixed address for NIC (optional).
--hint <key=value>	Hints for the scheduler (optional extension)
--config-drive <config-drive-volume>|True	Use specified volume as the config drive, or 'True' to use an ephemeral drive
--min <count>	Minimum number of servers to launch (default=1)
--max <count>	Maximum number of servers to launch (default=1)
--wait	Wait for build to complete

选项--image 用于指定为实例创建启动盘的镜像文件。选项--volume 用于指定为实例创建启动盘的卷（块设备）。这个卷必须基于一个云镜像来创建，例如，先创建一个基于 cirros 镜像的、大小为 2GB 的卷，执行如下命令。

```
openstack volume create --image cirros --size 2 --availablility-zone nova myvolume
```

再基于该卷创建实例，执行如下命令。

```
openstack server create --flavor m1.tiny   --volume myvolume   --nic net-id=public \
--security-group default   --key-name demo-key   cirros-20
```

这个实例本身保存在该启动卷上，重启不会丢失实例运行时添加或修改的数据。使用--volume 选项会自动创建一个启动索引号为 0 的块设备映射。在许多 Hypervisor 上，这个设备就是 vda。

6.3 项目实施

6.3.1 基于 Web 界面管理计算服务

基于部署远程桌面管理 OpenStack 平台，可以验证和操作计算服务。用户以云管理员身份登录 Dashboard 界面，可以执行计算服务管理操作。

1. 查看云主机类型

在 Dashboard 界面中，单击"管理员"主节点，再单击"系统"→"云主机类型"子节点，打开云主机类型列表界面，可以查看当前所有云主机类型，如图 6.10 所示。

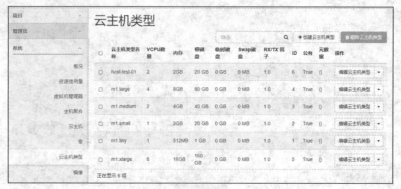

图 6.10　查看当前所有云主机类型

2. 创建云主机类型

在云主机类型列表界面的右上方单击"+创建云主机类型"按钮,打开"创建云主机类型"对话框,如图 6.11 所示,设置云主机类型的相关信息,如云主机类型名称、ID、vCPU 数量、内存(单位为 MB)、根磁盘(单位为 GB)、临时磁盘(单位为 GB)、Swap 磁盘(单位为 MB)、RX/TX 因子等。其中,带"*"的为必填项,对于带"?"的,将鼠标指针放到上面,可以显示相关提示信息。例如,将鼠标指针放到 ID 上,将显示相关提示信息,如图 6.12 所示。

图 6.11 "创建云主机类型"对话框

图 6.12 相关提示信息

在"创建云主机类型"对话框中,选择"云主机类型使用权"选项卡,可以选择该云主机类型适用的项目。如果不选择任何项目,则表示云主机可以在所有项目中使用。这里选择"admin"选项,单击"admin"右侧的"+"按钮,将其添加到选中的项目的列表中,如图 6.13 所示。

单击"创建云主机类型"按钮,创建云主机类型,返回云主机类型列表界面,可以查看创建的云主机类型,如图 6.14 所示。

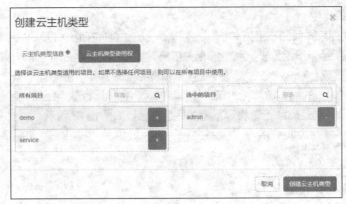

图 6.13 “云主机类型使用权”选项卡

云主机类型名称	VCPU数量	内存	根磁盘	临时磁盘	Swap磁盘	RX/TX 因子	ID	公有	元数据	操作
host-test-01	2	2GB	20 GB	0 GB	0 MB	1.0	6	True	{}	编辑云主机类型
m1.large	4	8GB	80 GB	0 GB	0 MB	1.0	4	True	{}	编辑云主机类型
m1.medium	2	4GB	40 GB	0 GB	0 MB	1.0	3	True	{}	编辑云主机类型
m1.small	1	2GB	20 GB	0 GB	0 MB	1.0	2	True	{}	编辑云主机类型
m1.tiny	1	512MB	1 GB	0 GB	0 MB	1.0	1	True	{}	编辑云主机类型
m1.xlarge	8	16GB	160 GB	0 GB	0 MB	1.0	5	True	{}	编辑云主机类型
yun_host-01	2	2GB	20 GB	0 GB	0 MB	1.0	bbda0c00-0a86-46f9-864e-f0aa75d53568	False	{}	编辑云主机类型

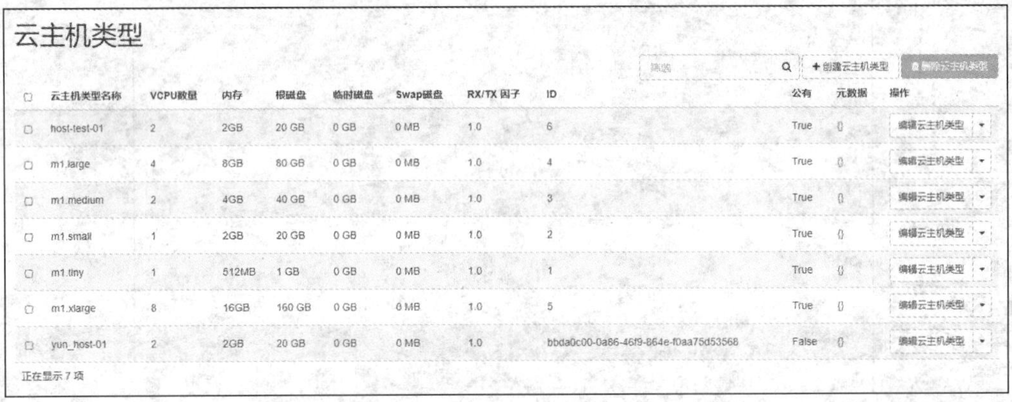

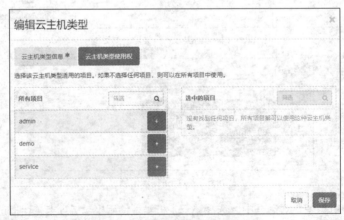

图 6.14　查看创建的云主机类型

3. 编辑云主机类型

在云主机类型列表界面中，选择要编辑的云计算类型，单击右侧的“编辑云主机类型”按钮，打开“编辑云主机类型”对话框，如图 6.15 所示，可以进行相关编辑操作。

图 6.15 “编辑云主机类型”对话框

4. 删除云主机类型

在云主机类型列表界面中，选择要删除的云计算类型，再单击右侧的“删除云主机类型”按钮或选择“操作”下拉列表中的“删除云主机类型”选项，如图 6.16 所示。

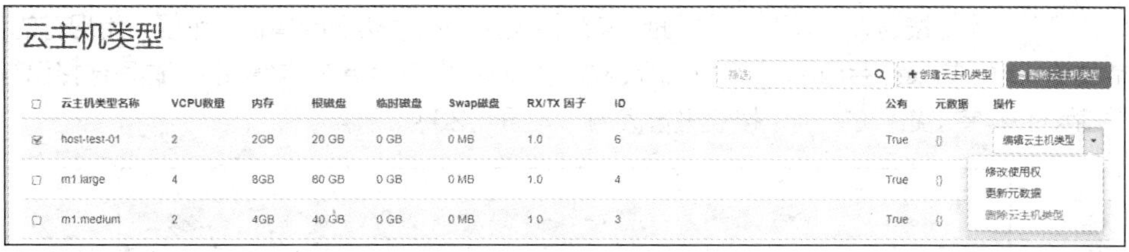

图 6.16　删除云主机类型

打开"确认删除云主机类型"对话框，如图 6.17 所示。

图 6.17　"确认删除云主机类型"对话框

5. 查看云主机

在 Dashboard 界面中，单击"项目"主节点，再单击"计算"→"云主机"子节点，打开云主机列表界面，可以查看当前所有云主机，如图 6.18 所示。

图 6.18　查看当前所有云主机

6. 创建云主机

在云主机列表界面的右上方单击"创建云主机"按钮，打开"启动实例"对话框，在"详细信息"界面中可进行详细信息设置，如图 6.19 所示。可以提供实例的初始主机名、将要部署的可用区域和实例计数，增加计数以创建具有多个同样设置的实例。

图 6.19　详细信息设置

单击"下一步"按钮，打开"源"界面，实例源是用来创建实例的模板。可以使用现有的实例、映像或卷的快照（如果已启用），也可以选择通过创建新卷来使用持久的存储器。这里选择已有的镜像源"cirros"，单击右侧的"+"按钮进行添加，如图 6.20 所示。

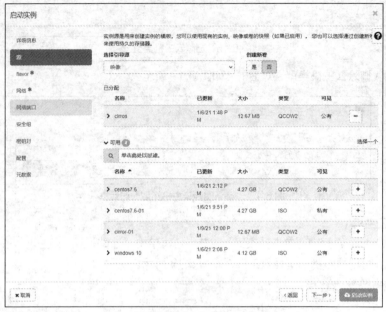

图 6.20　源设置

单击"下一步"按钮，打开"flavor"界面。flavor 用于管理实例的计算、内存和存储容量的大小。选择一个可用的云主机实例，单击右侧的"+"按钮进行添加，如图 6.21 所示。

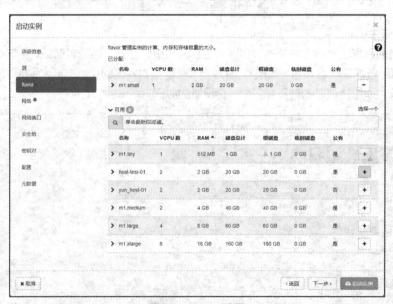

图 6.21　flavor 设置

单击"下一步"按钮，打开"网络"界面，在云中，网络为实例提供通信渠道。这里选择内部网络，单击右侧的"+"按钮进行添加，如图 6.22 所示。

图 6.22　网络设置

单击"下一步"按钮,打开"网络端口"界面,端口为实例提供了额外的通信渠道。可以选择端口而非网络或者二者都选择,如图 6.23 所示。

图 6.23　网络端口设置

单击"下一步"按钮,打开"安全组"界面,要在其中启动实例的安全组,如图 6.24 所示。

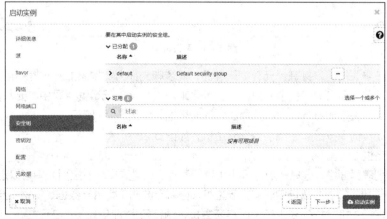

图 6.24　安全组设置

单击"下一步"按钮,打开"密钥对"界面,密钥对允许 SSH 到新创建的实例。可以选择一个现有的密钥对、导入一个密钥对或生成一个新的密钥对,如图 6.25 所示。

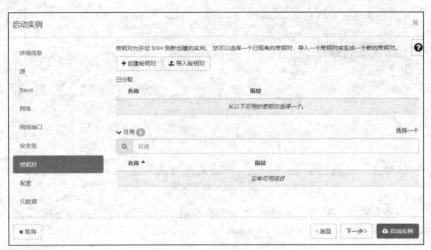

图 6.25 密钥对设置

单击"下一步"按钮,打开"配置"界面,使用此处可用的选项启动实例后,就可以定制该实例,如图 6.26 所示。其中,"自定义脚本"与其他系统中的"用户数据"类似。

图 6.26 配置设置

单击"下一步"按钮,打开"元数据"界面,此步骤允许为实例添加元数据条目。可以通过把左列的项目移动到右列来指定资源元数据。左列是来自 Glance 元数据目录中的元数据定义。请使用"自定义"选项来添加具有自己所选择的键的元数据,如图 6.27 所示。

单击"启动实例"按钮,即成功创建云主机,如图 6.28 所示。

7. 管理云主机

在云主机列表界面中,选择相应的云主机进行管理,在"操作"下拉列表中可以选择创建快照、绑定浮动 IP、连接接口、分离接口、编辑云主机、更新元数据、编辑元数据、编辑安全组、控制台、查看日志、暂停云主机、挂起云主机、废弃云主机、调整云主机大小、锁定云主机、解锁云主机、软重启云主机、硬重启云主机、关闭云主机、重建云主机、删除云主机等相关操作,如图 6.29 所示。

图 6.27 元数据设置

图 6.28 成功创建云主机

图 6.29 管理云主机

（1）创建快照

在云主机列表界面中，选择相应的云主机，选择"操作"下拉列表中的"创建快照"选项，打开"创建快照"对话框，如图 6.30 所示，输入快照名称，单击"创建快照"按钮即可。

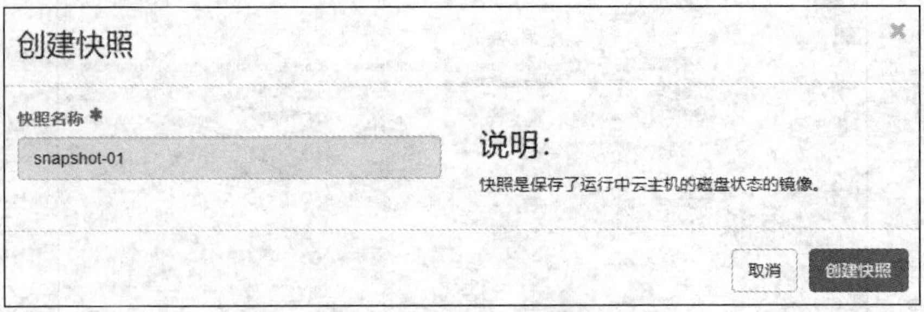

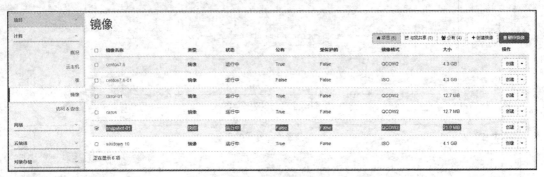

图 6.30 "创建快照"对话框

查看创建快照生成的镜像，在 Dashboard 界面中，单击"项目"主节点，再单击"计算"→"镜像"子节点，可以查看刚刚生成的快照镜像，如图 6.31 所示。

图 6.31 快照镜像

（2）绑定浮动 IP 地址

在云主机列表界面中，选择相应的云主机，选择"操作"下拉列表中的"绑定浮动 IP"选项，打开"管理浮动 IP 的关联"对话框，如图 6.32 所示。

图 6.32 "管理浮动 IP 的关联"对话框

在"选择一个 IP 地址"选项框的右侧，单击"+"按钮，打开"分配浮动 IP"对话框，如图 6.33 所示。

单击"分配 IP"按钮，打开"管理浮动 IP 的关联"对话框，完成浮动 IP 地址的关联，如图 6.34 所示。

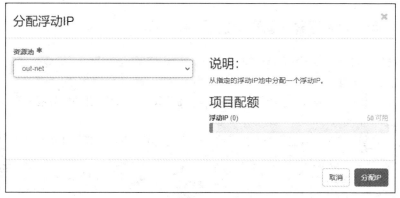

图 6.33 "分配浮动 IP"对话框

图 6.34 完成浮动 IP 地址的关联

单击"关联"按钮,返回云主机列表界面,即可查看浮动 IP 地址关联情况,如图 6.35 所示。

图 6.35 查看浮动 IP 地址关联情况

在云主机列表界面中,选择相应的云主机,选择"操作"下拉列表中的"解除浮动 IP 的绑定"选项,可以解除浮动 IP 地址的绑定,如图 6.36 所示。

图 6.36 解除浮动 IP 地址的绑定

6.3.2 基于命令行界面管理计算服务

系统管理员一般会基于命令行创建虚拟机实例,建议使用 openstack 命令来代替 nova 命令。

1. 云主机类型管理

可以进行查看云主机类型、查看云主机类型详细信息、创建云主机类型等相关操作。

(1)查看云主机类型

创建云主机,首先要设置云主机类型。查看云主机类型可以使用 openstack 命令,也可以使用 nova 命令,命令格式如下。

```
#openstack flavor list
```

或者

```
#nova flavor-list
```

V6-3 云主机类型管理

其命令的结果如图 6.37 所示。

图 6.37 查看云主机类型

(2)查看云主机类型详细信息

查看云主机类型详细信息可以使用 openstack 命令,也可以使用 nova 命令,命令格式如下。

```
# openstack flavor show 云主机类型名称或 ID
```

或者

```
# nova flavor-show 云主机类型名称或 ID
```

其命令的结果如图 6.38 所示。

图 6.38 查看云主机类型详细信息

（3）创建云主机类型

创建云主机类型可以使用 openstack 命令，也可以使用 nova 命令，执行如下命令，可以进行帮助查询。

```
# openstack help flavor create
```

或者

```
# nova help flavor-create
```

其命令的结果如下。

```
[root@controller ~]# openstack help flavor create
usage: openstack flavor create [-h]
                               [-f {html,json,json,shell,table,value,yaml,yaml}]
                               [-c COLUMN] [--max-width <integer>]
                               [--noindent] [--prefix PREFIX] [--id <id>]
                               [--ram <size-mb>] [--disk <size-gb>]
                               [--ephemeral <size-gb>] [--swap <size-gb>]
                               [--vcpus <vcpus>] [--rxtx-factor <factor>]
                               [--public | --private]
                               <flavor-name>

Create new flavor
positional arguments:
  <flavor-name>         New flavor name
optional arguments:
  -h, --help            show this help message and exit
  --id <id>             Unique flavor ID; 'auto' creates a UUID (default:
                        auto)
  --ram <size-mb>       Memory size in MB (default 256M)
  --disk <size-gb>      Disk size in GB (default 0G)
  --ephemeral <size-gb>
                        Ephemeral disk size in GB (default 0G)
  --swap <size-gb>      Swap space size in GB (default 0G)
  --vcpus <vcpus>       Number of vcpus (default 1)
  --rxtx-factor <factor>
                        RX/TX factor (default 1)
  --public              Flavor is available to other projects (default)
  --private             Flavor is not available to other projects
output formatters:
  output formatter options
  -f {html,json,json,shell,table,value,yaml,yaml},--format {html,json,json,shell,table,value,yaml,yaml}
                        the output format, defaults to table
  -c COLUMN, --column COLUMN
                        specify the column(s) to include, can be repeated
table formatter:
  --max-width <integer>
                        Maximum display width, 0 to disable
json formatter:
  --noindent            whether to disable indenting the JSON
shell formatter:
```

```
      a format a UNIX shell can parse (variable="value")
  --prefix PREFIX         add a prefix to all variable names
[root@controller ~]#
```

或者

```
[root@controller ~]# nova help flavor-create
usage: nova flavor-create [--ephemeral <ephemeral>] [--swap <swap>]
                          [--rxtx-factor <factor>] [--is-public <is-public>]
                          <name> <id> <ram> <disk> <vcpus>
Create a new flavor.
Positional arguments:
  <name>                  Unique name of the new flavor.
  <id>                    Unique ID of the new flavor. Specifying 'auto' will
                          generated a UUID for the ID.
  <ram>                   Memory size in MB.
  <disk>                  Disk size in GB.
  <vcpus>                 Number of vcpus
Optional arguments:
  --ephemeral <ephemeral> Ephemeral space size in GB (default 0).
  --swap <swap>           Swap space size in MB (default 0).
  --rxtx-factor <factor>  RX/TX factor (default 1).
  --is-public <is-public> Make flavor accessible to the public (default
                          true).
[root@controller ~]#
```

例如，使用 openstack 命令创建一个名称为 yun_host-10、ID 为 10、内存为 2048 MB、磁盘为 20GB、vCPU 数量为 2 的云主机类型，执行命令如下。

```
# openstack flavor create --id=10 --ram=2048 --disk=20 --vcpus=2 yun_host-10
```

其命令的结果如图 6.39 所示。

```
[root@controller ~]# openstack flavor create --id=10 --ram=2048 --disk=20 --vcpus=2 yun_host-10
+----------------------------+-------------+
| Field                      | Value       |
+----------------------------+-------------+
| OS-FLV-DISABLED:disabled   | False       |
| OS-FLV-EXT-DATA:ephemeral  | 0           |
| disk                       | 20          |
| id                         | 10          |
| name                       | yun_host-10 |
| os-flavor-access:is_public | True        |
| ram                        | 2048        |
| rxtx_factor                | 1.0         |
| swap                       |             |
| vcpus                      | 2           |
+----------------------------+-------------+
[root@controller ~]#
```

图 6.39　创建云主机类型 yun_host-10

例如，使用 nova 命令创建一个名为 yun_host-20、ID 为 11、内存为 2048MB、磁盘为 20GB、vCPU 数量为 2 的云主机类型，执行命令如下。

```
#nova flavor-create yun_host-20 11 2048 20 2
```

其命令的结果如图 6.40 所示。

```
[root@controller ~]# nova flavor-create yun_host-20 11 2048 20 2
+----+-------------+-----------+------+-----------+------+-------+-------------+-----------+
| ID | Name        | Memory_MB | Disk | Ephemeral | Swap | VCPUs | RXTX_Factor | Is_Public |
+----+-------------+-----------+------+-----------+------+-------+-------------+-----------+
| 11 | yun_host-20 | 2048      | 20   | 0         |      | 2     | 1.0         | True      |
+----+-------------+-----------+------+-----------+------+-------+-------------+-----------+
[root@controller ~]#
```

图 6.40　创建云主机类型 yun_host-20

查看创建的 yun_host-10 和 yun_host-20 云主机类型列表信息，如图 6.41 所示。

```
[root@controller ~]# openstack flavor list
+--------------------------------------+--------------+-------+------+-----------+-------+-----------+
| ID                                   | Name         |   RAM | Disk | Ephemeral | VCPUs | Is Public |
+--------------------------------------+--------------+-------+------+-----------+-------+-----------+
| 1                                    | m1.tiny      |   512 |    1 |         0 |     1 | True      |
| 10                                   | yun_host-10  |  2048 |   20 |         0 |     2 | True      |
| 11                                   | yun_host-20  |  2048 |   20 |         0 |     2 | True      |
| 2                                    | m1.small     |  2048 |   20 |         0 |     1 | True      |
| 3                                    | m1.medium    |  4096 |   40 |         0 |     2 | True      |
| 4                                    | m1.large     |  8192 |   80 |         0 |     4 | True      |
| 5                                    | m1.xlarge    | 16384 |  160 |         0 |     8 | True      |
| 6                                    | host-test-01 |  2048 |   20 |         0 |     2 | True      |
| bbda0c00-0a86-46f9-864e-f0aa75d53568 | yun_host-01  |  2048 |   20 |         0 |     2 | False     |
+--------------------------------------+--------------+-------+------+-----------+-------+-----------+
[root@controller ~]#
```

图 6.41　查看创建的云主机类型列表信息

2. 云主机管理

可以进行查看云主机列表信息、创建云主机、访问云主机等相关操作。

（1）查看云主机列表信息

查看云主机列表信息可以使用 openstack 命令，也可以使用 nova 命令，执行命令如下。

V6-4　云主机管理

```
# openstack  server  list
```

或者

```
# nova  list
```

其命令的结果如图 6.42 所示。

```
[root@controller ~]# openstack  server list
+--------------------------------------+---------------+--------+-------------------------------------------+
| ID                                   | Name          | Status | Networks                                  |
+--------------------------------------+---------------+--------+-------------------------------------------+
| 5282ca6c-f232-45e8-8fee-5cd94ab6fcda | cloud_host-01 | ACTIVE | in-net=192.168.100.101, 202.199.184.102 |
+--------------------------------------+---------------+--------+-------------------------------------------+
[root@controller ~]# nova list
+--------------------------------------+---------------+--------+------------+-------------+-------------------------------------------+
| ID                                   | Name          | Status | Task State | Power State | Networks                                  |
+--------------------------------------+---------------+--------+------------+-------------+-------------------------------------------+
| 5282ca6c-f232-45e8-8fee-5cd94ab6fcda | cloud_host-01 | ACTIVE | -          | Running     | in-net=192.168.100.101, 202.199.184.102 |
+--------------------------------------+---------------+--------+------------+-------------+-------------------------------------------+
[root@controller ~]#
```

图 6.42　查看云主机列表信息

（2）创建云主机

使用命令行创建云主机时，首先需要查看云主机相关资源，例如，云类型、网络、镜像等相关必需资源，可执行如下命令。

```
# nova flavor-list
# neutron net-list
#glance image-list
```

其命令的结果如图 6.43 所示。

```
[root@controller ~]# nova flavor-list
+--------------------------------------+--------------+-----------+------+-----------+------+-------+-------------+-----------+
| ID                                   | Name         | Memory_MB | Disk | Ephemeral | Swap | VCPUs | RXTX_Factor | Is_Public |
+--------------------------------------+--------------+-----------+------+-----------+------+-------+-------------+-----------+
| 1                                    | m1.tiny      | 512       | 1    | 0         |      | 1     | 1.0         | True      |
| 10                                   | yun_host-10  | 2048      | 20   | 0         |      | 2     | 1.0         | True      |
| 11                                   | yun_host-20  | 2048      | 20   | 0         |      | 2     | 1.0         | True      |
| 2                                    | m1.small     | 2048      | 20   | 0         |      | 1     | 1.0         | True      |
| 3                                    | m1.medium    | 4096      | 40   | 0         |      | 2     | 1.0         | True      |
| 4                                    | m1.large     | 8192      | 80   | 0         |      | 4     | 1.0         | True      |
| 5                                    | m1.xlarge    | 16384     | 160  | 0         |      | 8     | 1.0         | True      |
| 6                                    | host-test-01 | 2048      | 20   | 0         |      | 2     | 1.0         | True      |
| bbda0c00-0a86-46f9-864e-f0aa75d53568 | yun_host-01  | 2048      | 20   | 0         |      | 2     | 1.0         | False     |
+--------------------------------------+--------------+-----------+------+-----------+------+-------+-------------+-----------+
[root@controller ~]# neutron net-list
+--------------------------------------+---------+-------------------------------------------------------+
| id                                   | name    | subnets                                               |
+--------------------------------------+---------+-------------------------------------------------------+
| ce4678aa-a2c7-4399-a215-6907d929eaa5 | in-net  | 83cb607b-8e56-4d8f-b7b5-d84c375d660a 192.168.100.0/24 |
| afddc555-61cb-4cf8-9bd4-1316079e6355 | out-net | 4207b78f-f5ab-4295-af9b-ae2a1ebe8032 202.199.184.0/24 |
+--------------------------------------+---------+-------------------------------------------------------+
[root@controller ~]# glance image-list
+--------------------------------------+---------------+
| ID                                   | Name          |
+--------------------------------------+---------------+
| 8e6e2298-bc00-4033-8ee4-3eb126053483 | centos7.6     |
| f7b0213f-8edd-441f-b217-ecfb6754b297 | centos7.6-01  |
| 29ca2fc6-89d4-4771-987b-9d0fc405cc04 | cirror-01     |
| b422d06f-b8d0-49ee-b14c-62df4731c0d1 | cirros        |
| c0419648-e8a4-4e13-aa9e-77a5e21e6ce1 | snapshot-01   |
| 9ac41140-727c-4c52-a7bb-6880016a4ab2 | windows 10    |
+--------------------------------------+---------------+
[root@controller ~]#
```

图 6.43　查看云主机相关资源

创建云主机可以使用 openstack 命令，也可以使用 nova 命令，执行如下命令，可以进行帮助查询。

```
# openstack help server create
```

或者

```
# nova help boot
```

其命令的结果如下。

```
[root@controller ~]# openstack help server create
usage: openstack server create [-h]
                               [-f {html,json,json,shell,table,value,yaml,yaml}]
                               [-c COLUMN] [--max-width <integer>]
                               [--noindent] [--prefix PREFIX]
                               (--image <image> | --volume <volume>) --flavor
                               <flavor>
                               [--security-group <security-group-name>]
                               [--key-name <key-name>]
                               [--property <key=value>]
                               [--file <dest-filename=source-filename>]
                               [--user-data <user-data>]
                               [--availability-zone <zone-name>]
                               [--block-device-mapping <dev-name=mapping>]
                               [--nic <net-id=net-uuid,v4-fixed-ip=ip-addr,v6-fixed-ip=ip-addr,port-id=port-uuid>]
                               [--hint <key=value>]
                               [--config-drive <config-drive-volume>|True]
                               [--min <count>] [--max <count>] [--wait]
                               <server-name>
[root@controller ~]#
```

或者

```
[root@controller ~]# nova help boot
usage: nova boot [--flavor <flavor>] [--image <image>]
                 [--image-with <key=value>] [--boot-volume <volume_id>]
                 [--snapshot <snapshot_id>] [--min-count <number>]
                 [--max-count <number>] [--meta <key=value>]
                 [--file <dst-path=src-path>] [--key-name <key-name>]
                 [--user-data <user-data>]
                 [--availability-zone <availability-zone>]
                 [--security-groups <security-groups>]
                 [--block-device-mapping <dev-name=mapping>]
                 [--block-device key1=value1[,key2=value2...]]
                 [--swap <swap_size>]
                 [--ephemeral size=<size>[,format=<format>]]
                 [--hint <key=value>]
                 [--nic <net-id=net-uuid,net-name=network-name,v4-fixed-ip=ip-addr,v6-fixed-ip=ip-addr,port-id=port-uuid>]
                 [--config-drive <value>] [--poll] [--admin-pass <value>]
                 [--access-ip-v4 <value>] [--access-ip-v6 <value>]
                 [--description <description>]
```

```
               <name>
[root@controller ~]#
```

例如，使用 openstack 命令创建一个云主机，云主机类型为 yun_host-10、镜像为 cirros、网络 ID 为内部网络 ID（不是内部网络子网 ID）、云主机名为 cloud_host-10，执行命令如下。

```
[root@controller ~]# openstack server create --flavor yun_host-10 --image cirros  --nic net-id=ce4678aa-a2c7-4399-a215-6907d929eaa5 cloud_host-10
```

其命令的结果如图 6.44 所示。

```
[root@controller ~]#  openstack server create --flavor yun_host-10 --image cirros  --nic net-id=ce4678aa-a2c7-4399-a215-6907d929eaa5 cloud_host-10
+-------------------------------------+-----------------------------------------------+
| Field                               | Value                                         |
+-------------------------------------+-----------------------------------------------+
| OS-DCF:diskConfig                   | MANUAL                                        |
| OS-EXT-AZ:availability_zone         |                                               |
| OS-EXT-SRV-ATTR:host                | None                                          |
| OS-EXT-SRV-ATTR:hypervisor_hostname | None                                          |
| OS-EXT-SRV-ATTR:instance_name       | instance-00000003                             |
| OS-EXT-STS:power_state              | 0                                             |
| OS-EXT-STS:task_state               | scheduling                                    |
| OS-EXT-STS:vm_state                 | building                                      |
| OS-SRV-USG:launched_at              | None                                          |
| OS-SRV-USG:terminated_at            | None                                          |
| accessIPv4                          |                                               |
| accessIPv6                          |                                               |
| addresses                           |                                               |
| adminPass                           | f2pFXojMVYyh                                  |
| config_drive                        |                                               |
| created                             | 2021-01-11T22:33:10Z                          |
| flavor                              | yun_host-10 (10)                              |
| hostId                              |                                               |
| id                                  | 6e9402bf-9f1c-402b-bd6c-8ec711af8c5d          |
| image                               | cirros (b422d06f-b8d0-49ee-b14c-62df4731c0d1) |
| key_name                            | None                                          |
| name                                | cloud_host-10                                 |
| os-extended-volumes:volumes_attached | []                                           |
| progress                            | 0                                             |
| project_id                          | f9ff39ba9daa4e5a8fee1fc50e2d2b34              |
| properties                          |                                               |
| security_groups                     | [{u'name': u'default'}]                       |
| status                              | BUILD                                         |
| updated                             | 2021-01-11T22:33:11Z                          |
| user_id                             | 0befa70f767848e39df8224107b71858              |
+-------------------------------------+-----------------------------------------------+
[root@controller ~]#
```

图 6.44　使用 openstack 命令创建云主机

例如，使用 nova 命令创建一个云主机，云主机类型为 yun_host-20、镜像为 cirror-01、网络 ID 为外部网络 ID（不是外部网络子网 ID）、云主机名为 cloud_host-20，执行命令如下。

```
[root@controller ~]# nova  boot  --flavor  yun_host-20  --image cirror-01  --nic net-id=afddc555-61cb-4cf8-9bd4-1316079e6355 cloud_host-20
```

其命令的结果如图 6.45 所示。

```
[root@controller ~]# nova boot --flavor yun_host-20  --image cirror-01   --nic net-id=afddc555-61cb-4cf8-9bd4-1316079e6355 cloud_host-20
+--------------------------------------+--------------------------------------------------+
| Property                             | Value                                            |
+--------------------------------------+--------------------------------------------------+
| OS-DCF:diskConfig                    | MANUAL                                           |
| OS-EXT-AZ:availability_zone          |                                                  |
| OS-EXT-SRV-ATTR:host                 | -                                                |
| OS-EXT-SRV-ATTR:hostname             | cloud-host-20                                    |
| OS-EXT-SRV-ATTR:hypervisor_hostname  | -                                                |
| OS-EXT-SRV-ATTR:instance_name        | instance-00000004                                |
| OS-EXT-SRV-ATTR:kernel_id            |                                                  |
| OS-EXT-SRV-ATTR:launch_index         | 0                                                |
| OS-EXT-SRV-ATTR:ramdisk_id           |                                                  |
| OS-EXT-SRV-ATTR:reservation_id       | r-37k2hbjc                                       |
| OS-EXT-SRV-ATTR:root_device_name     | -                                                |
| OS-EXT-SRV-ATTR:user_data            | -                                                |
| OS-EXT-STS:power_state               | 0                                                |
| OS-EXT-STS:task_state                | scheduling                                       |
| OS-EXT-STS:vm_state                  | building                                         |
| OS-SRV-USG:launched_at               | -                                                |
| OS-SRV-USG:terminated_at             | -                                                |
| accessIPv4                           |                                                  |
| accessIPv6                           |                                                  |
| adminPass                            | Pi4y378Rsrf3                                     |
| config_drive                         |                                                  |
| created                              | 2021-01-11T22:41:48Z                             |
| description                          | -                                                |
| flavor                               | yun_host-20 (11)                                 |
| hostId                               |                                                  |
| host_status                          |                                                  |
| id                                   | 27a462e8-60ef-4d1b-87dd-5a63a22fb300             |
| image                                | cirror-01 (29ca2fc6-89d4-4771-987b-9d0fc405cc04) |
| key_name                             | -                                                |
| locked                               | False                                            |
| metadata                             | {}                                               |
| name                                 | cloud_host-20                                    |
| os-extended-volumes:volumes_attached | []                                               |
| progress                             | 0                                                |
| security_groups                      | default                                          |
| status                               | BUILD                                            |
| tenant_id                            | f9ff39ba9daa4e5a8fee1fc50e2d2b34                 |
| updated                              | 2021-01-11T22:41:48Z                             |
| user_id                              | 0befa70f767848e39df8224107b71858                 |
+--------------------------------------+--------------------------------------------------+
[root@controller ~]#
```

图 6.45　使用 nova 命令创建云主机

使用命令行查看云主机资源，如图 6.46 所示。

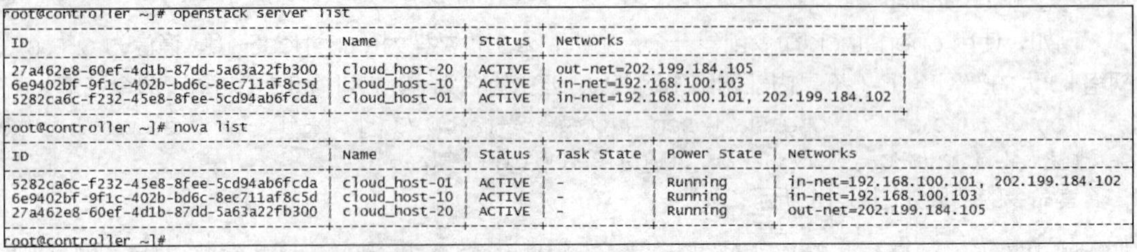

```
[root@controller ~]# openstack server list
+--------------------------------------+---------------+--------+-----------------------------------------+
| ID                                   | Name          | Status | Networks                                |
+--------------------------------------+---------------+--------+-----------------------------------------+
| 27a462e8-60ef-4d1b-87dd-5a63a22fb300 | cloud_host-20 | ACTIVE | out-net=202.199.184.105                 |
| 6e9402bf-9f1c-402b-bd6c-8ec711af8c5d | cloud_host-10 | ACTIVE | in-net=192.168.100.103                  |
| 5282ca6c-f232-45e8-8fee-5cd94ab6fcda | cloud_host-01 | ACTIVE | in-net=192.168.100.101, 202.199.184.102 |
+--------------------------------------+---------------+--------+-----------------------------------------+
[root@controller ~]# nova list
+--------------------------------------+---------------+--------+------------+-------------+-----------------------------------------+
| ID                                   | Name          | Status | Task State | Power State | Networks                                |
+--------------------------------------+---------------+--------+------------+-------------+-----------------------------------------+
| 5282ca6c-f232-45e8-8fee-5cd94ab6fcda | cloud_host-01 | ACTIVE | -          | Running     | in-net=192.168.100.101, 202.199.184.102 |
| 6e9402bf-9f1c-402b-bd6c-8ec711af8c5d | cloud_host-10 | ACTIVE | -          | Running     | in-net=192.168.100.103                  |
| 27a462e8-60ef-4d1b-87dd-5a63a22fb300 | cloud_host-20 | ACTIVE | -          | Running     | out-net=202.199.184.105                 |
+--------------------------------------+---------------+--------+------------+-------------+-----------------------------------------+
[root@controller ~]#
```

图 6.46 使用命令行查看云主机资源

在 Dashboard 界面中，单击"项目"主节点，再单击"计算"→"云主机"子节点，打开云主机列表界面，如图 6.47 所示，可以查看当前所有云主机资源。

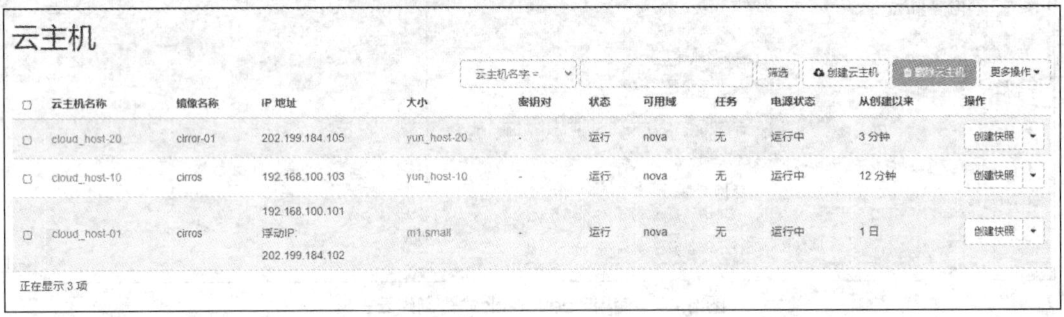

云主机名称	镜像名称	IP 地址	大小	密钥对	状态	可用域	任务	电源状态	从创建以来	操作
cloud_host-20	cirror-01	202.199.184.105	yun_host-20	-	运行	nova	无	运行中	3 分钟	创建快照
cloud_host-10	cirros	192.168.100.103	yun_host-10	-	运行	nova	无	运行中	12 分钟	创建快照
cloud_host-01	cirros	192.168.100.101 浮动IP: 202.199.184.102	m1.small	-	运行	nova	无	运行中	1 日	创建快照

正在显示 3 项

图 6.47 使用 Web 界面查看云主机资源

（3）访问云主机

可以通过 Web 浏览器方式访问云主机，获得 Web 浏览器的 URL 信息。可通过帮助命令执行相关命令参数，执行命令如下。

```
[root@controller ~]# nova help | grep vnc
    get-vnc-console             Get a vnc console to a server.
[root@controller ~]# nova help get-vnc-console
usage: nova get-vnc-console <server> <console-type>
Get a vnc console to a server.
Positional arguments:
  <server>          Name or ID of server.
  <console-type>    Type of vnc console ("novnc" or "xvpvnc").
[root@controller ~]#
```

其命令的结果如图 6.48 所示。

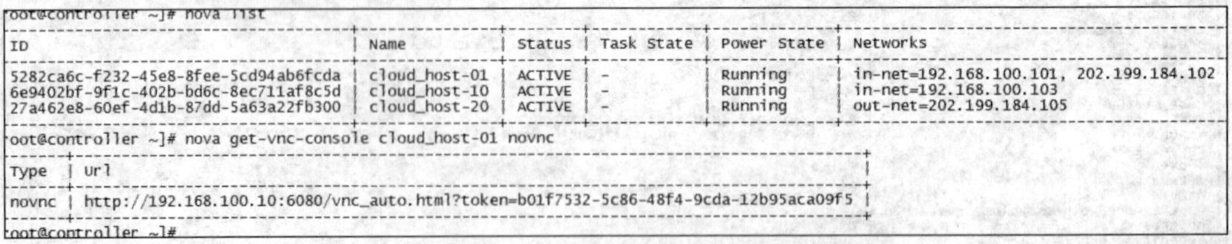

```
[root@controller ~]# nova list
+--------------------------------------+---------------+--------+------------+-------------+-----------------------------------------+
| ID                                   | Name          | Status | Task State | Power State | Networks                                |
+--------------------------------------+---------------+--------+------------+-------------+-----------------------------------------+
| 5282ca6c-f232-45e8-8fee-5cd94ab6fcda | cloud_host-01 | ACTIVE | -          | Running     | in-net=192.168.100.101, 202.199.184.102 |
| 6e9402bf-9f1c-402b-bd6c-8ec711af8c5d | cloud_host-10 | ACTIVE | -          | Running     | in-net=192.168.100.103                  |
| 27a462e8-60ef-4d1b-87dd-5a63a22fb300 | cloud_host-20 | ACTIVE | -          | Running     | out-net=202.199.184.105                 |
+--------------------------------------+---------------+--------+------------+-------------+-----------------------------------------+
[root@controller ~]# nova get-vnc-console cloud_host-01 novnc
+-------+--------------------------------------------------------------------------------+
| Type  | Url                                                                            |
+-------+--------------------------------------------------------------------------------+
| novnc | http://192.168.100.10:6080/vnc_auto.html?token=b01f7532-5c86-48f4-9cda-12b95aca09f5 |
+-------+--------------------------------------------------------------------------------+
[root@controller ~]#
```

图 6.48 获得 Web 浏览器的 URL 信息

在 Web 浏览器的地址栏中，输入获得的 URL 信息，可以访问云主机 cloud-host-01，查看云主机资源，如图 6.49 所示，输入用户名和密码进行登录，用户名为 cirros，密码为 cubswin:)。

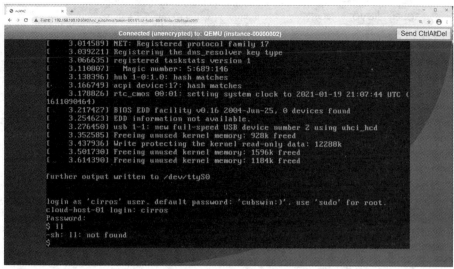

图 6.49 查看云主机资源

课后习题

1. 选择题

（1）Nova 支持通过多种调度方式来选择运行虚拟机的主机节点，默认使用的调度器是（　　）。
　　A. 随机调度器　　　　　　　　　　B. 过滤器调度器
　　C. 缓存调度器　　　　　　　　　　D. 固定调度器

（2）过滤调度器需要执行调度操作时，默认的第一个过滤器是（　　）。
　　A. 再审过滤器　　B. 可用区域过滤器　　C. 内存过滤器　　D. 磁盘过滤器

（3）根据可用磁盘空间来调度虚拟机创建时，通过 nova.conf 中的 disk_allocation_ratio 参数控制，其默认值为（　　）。
　　A. 0.5　　　　　B. 1.0　　　　　C. 1.5　　　　　D. 2.0

（4）根据可用 CPU 核心来调度虚拟机创建时，通过 nova.conf 中的 cpu_allocation_ratio 参数控制，其默认值为（　　）。
　　A. 10　　　　　B. 15　　　　　C. 16　　　　　D. 20

（5）【多选】查看云主机类型的命令是（　　）。
　　A. openstack flavor list　　　　　B. openstack flavor-list
　　C. nova flavor list　　　　　　　D. nova flavor-list

（6）【多选】查看云主机类型详细信息的命令是（　　）。
　　A. openstack flavor show 云主机类型名称或 ID
　　B. openstack flavor list 云主机类型名称或 ID
　　C. nova flavor-show 云主机类型名称或 ID
　　D. nova flavor-list 云主机类型名称或 ID

（7）【多选】创建云主机类型的命令是（　　）。
　　A. openstack flavor create　　　　B. openstack flavor set

　　　　　C. nova flavor-create　　　　　D. nova flavor-set
（8）【多选】查看云主机的命令是（　　）。
　　　　　A. openstack server list　　　　B. openstack server-list
　　　　　C. nova server-list　　　　　　D. nova list
（9）【多选】创建云主机的命令是（　　）。
　　　　　A. openstack server create　　　B. openstack server-create
　　　　　C. nova server-create　　　　　D. nova boot
（10）OpenStack 中负责计算的模块是（　　）。
　　　　　A. Keystone　　B. Glance　　C. Nova　　D. Cinder

2. 简答题

（1）简述 Nova 以及 Nova 的系统架构。
（2）简述 API 组件的作用。
（3）简述 Conductor 组件的作用。
（4）简述 Scheduler 组件的作用。
（5）简述 Compute 组件的作用
（6）简述过滤器调度器的调度过程。
（7）简述虚拟机实例化流程。
（8）简述 Nova 物理部署以及 Nova 的 Cell 架构。

项目7
OpenStack存储服务

【学习目标】

- 掌握Cinder块存储服务基础。
- 掌握Cinder的系统架构。
- 掌握Swift对象存储服务基础。
- 掌握Swift的系统架构。
- 掌握基于Web界面存储服务管理配置方法。
- 掌握基于命令行界面存储服务管理配置方法。

7.1 项目陈述

与网络一样,存储也是 OpenStack 最重要的基础设施之一。Nova 实现的虚拟机实例需要存储的支持,这些存储可分为临时性存储和持久性存储两种类型。在 OpenStack 项目中,通过 Nova 创建实例时可直接利用节点的本地存储为虚拟机提供临时性存储。这种存储空间主要作为虚拟机的根磁盘来运行操作系统,也可作为其他磁盘暂存数据,其大小由所用的实例类型决定。实例使用临时性存储来保存所有数据,一旦实例被关闭、重启或删除,该实例中的数据就会全部丢失。如果指定使用持久性存储,则可以保证这些数据不会丢失,使得数据持续可用,不受虚拟机实例终止的影响。当然,这种持久性并不是绝对的,一旦它本身被删除或损坏,其中的数据也会丢失。目前,OpenStack 提供持久性存储服务的项目有 Cinder 的块存储(Block Storage)、Swift 的对象存储(Object Storage)。块存储又称卷存储(Volume Storage),为用户提供基于数据块的存储设备访问,以卷的形式提供给虚拟机实例挂载,为实例提供额外的磁盘空间。对象存储所存放的数据通常称为对象(Object),实际上就是文件,可用于为虚拟机实例提供备份、归档的存储空间,包括虚拟机镜像的保存。

7.2 必备知识

7.2.1 Cinder 块存储服务基础

OpenStack 从 Folsom 版本开始将 Nova 中的持久性块存储功能组件 Nova-Volume 剥离出

来，独立为 OpenStack 块存储服务，并将其命名为 Cinder。与 Nova 利用主机本地存储为虚拟机提供的临时存储不同，Cinder 为虚拟机提供持久性的存储，并实现虚拟机存储卷的生命周期管理，因此又称卷存储服务。

Cinder 是块存储，可以把 Cinder 当作优秀管理程序来理解。Cinder 块存储具有安全可靠、高并发、大吞吐量、低时延、规格丰富、简单易用的特点，适用于文件系统、数据库或者其他需要原始块设备的系统软件或应用。

可以用 Cinder 创建卷，并将它连接到虚拟机上，这个卷就像虚拟机的一个存储分区一样工作。如果结束虚拟机的运行，则卷和其中的数据依然存在，可以把它连接到其他虚拟机上继续使用其中的数据。Cinder 创建的卷必须被连接到虚拟机上才能工作，可以把 Cinder 理解成一块可移动硬盘。

1. Cinder 的主要功能

Cinder 的核心功能是对卷的管理，允许对卷、卷的类型、卷的快照进行处理，它并没有实现对块设备的管理和实际服务，而是通过后端的统一存储接口支持不同块设备厂商的块存储服务，实现其驱动，支持与 OpenStack 进行整合。Cinder 提供的是一种存储基础设施服务，为用户提供基于数据块的存储设备访问，其具体功能如下。

V7-1 Cinder 的主要功能

（1）为管理块存储提供一套方法，对卷实现从创建到删除的整个生命周期管理，允许对卷、卷的类型、卷的快照进行处理。

（2）提供持久性块存储资源，供 Nova 计算服务的虚拟机实例使用。从实例的角度看，挂载的每个卷都是一块磁盘。使用 Cinder 可以将一个存储设备连接到一个实例。另外，可以将镜像写到块存储设备中，让 Nova 计算服务用作可启动的持久性实例。

（3）对不同的后端存储进行封装，对外提供统一的 API。

2. Cinder 的系统架构

V7-2 Cinder 的系统架构

Cinder 延续了 Nova 以及其他 OpenStack 组件的设计思想，其系统架构如图 7.1 所示，主要包括 cinder-api、cinder-scheduler、cinder-volume 和 cinder-backup 服务，这些服务之间通过消息队列协议进行通信。

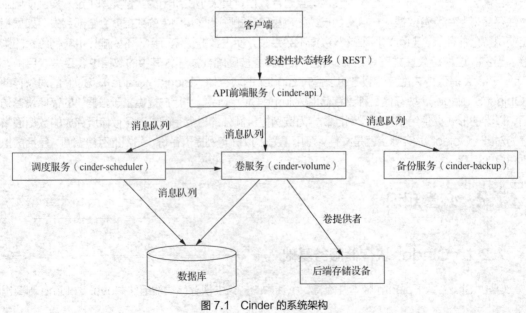

图 7.1 Cinder 的系统架构

（1）客户端

客户端可以是 OpenStack 的最终用户，也可以是其他程序，包括终端用户、命令行和 OpenStack 的其他组件，凡是 Cinder 服务提出请求的都是 Cinder 客户端。

（2）API 前端服务（cinder-api）

cinder-api 作为 Cinder 对外服务的 HTTP 接口，向客户端呈现 Cinder 能够提供的功能，负责接收和处理 REST 请求并将请求发送给消息队列（MQ 队列）。当客户需要执行卷的相关操作时，能且只能向 cinder-api 发送 REST 请求。

（3）调度服务（cinder-scheduler）

cinder-scheduler 对请求进行调度，将请求转发到合适的卷服务，即处理任务队列的任务，通过调度算法选择最合适的存储节点以创建卷。

（4）卷服务（cinder-volume）

调度服务只分配任务，真正执行任务的是卷服务。cinder-volume 管理块存储设备，定义后端设备。运行 cinder-volume 服务的节点被称为存储节点。

（5）备份服务（cinder-backup）

备份服务用于提供卷的备份功能，支持将块存储卷备份到 OpenStack 对象存储 Swift 中。

（6）数据库

Cinder 有一些数据需要存放到数据库中，一般使用 MySQL 数据库。数据库是安装在控制节点上的。

（7）卷提供者

块存储服务需要后端存储设备来创建卷，如外部的磁盘阵列以及其他存储设施等。卷提供者定义了存储设备，为卷提供物理存储空间。cinder-volume 支持多种卷提供者，每种卷提供者都能通过自己的驱动与 cinder-volume 协调工作。

（8）消息队列

Cinder 得到请求后会自动访问块存储服务，它有两个显著的特点：第一，用户必须提出请求，服务才会进行响应；第二，用户可以使用自定义的方式实现半自动化服务。简而言之，Cinder 虚拟化块存储设备，提供给用户自助服务的 API 用以请求和使用存储池中的资源，而 Cinder 本身并不能获取具体的存储形式和物理设备信息。

Cinder 的各个子服务通过消息队列实现进程间的通信和相互协作。有了消息队列，子服务之间实现了相互交流，这种松散的结构也是分布式系统的重要特征。

Cinder 创建卷的基本流程是客户端向 cinder-api 发送请求，要求创建一个卷；cinder-api 对请求做一些必要处理后，向 RabbitMQ 发送一条消息，让 cinder-scheduler 服务创建一个卷；cinder-scheduler 从消息队列中获取 cinder-api 发给它的消息，然后执行调度算法，从若干存储节点中选出某节点；cinder-scheduler 向消息队列发送一条消息，让该存储节点创建这个卷；该存储节点的 cinder-volume 服务从消息队列中获取 cinder-scheduler 发给它的消息，然后通过驱动在卷提供者定义的后端存储设备上创建卷。

3. Cinder 块存储服务与 Nova 计算服务之间的交互

Cinder 块存储服务与 Nova 计算服务进行交互，为虚拟机实例提供卷。Cinder 负责卷的全生命周期管理。Nova 的虚拟机实例通过连接 Cinder 的卷将该卷作为其存储设备，用户可以对其进行读写、格式化等操作。分离卷将使虚拟机不再使用它，但是该卷上的数据不受影响，数据依然保持完整，还可以连接到该虚拟机上或其他虚拟机上，如图 7.2 所示。

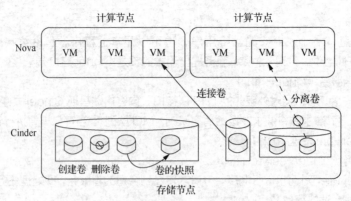

图 7.2 Cinder 块存储服务与 Nova 计算服务之间的交互

通过 Cinder 可以方便地管理虚拟机的存储，虚拟机的整个生命周期中对应的卷操作如图 7.3 所示。

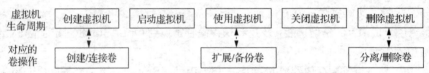

图 7.3 虚拟机的整个生命周期中对应的卷操作

4. cinder-api 服务

cinder-api 服务的主要功能是接收客户端发来的 HTTP 请求，在整个块存储服务中进行验证和路由请求。作为整个 Cinder 服务的门户，所有对 Cinder 的请求都先由它进行处理。目前，cinder-api 有 3 个版本，即 cinder、cinderv2 和 cinderv3。在 Keystone 中可以查看 cinder-api 的端点列表信息，执行如下命令。

```
# openstack  endpoint  list
```

其命令的结果如图 7.4 所示。

```
[root@controller ~]# openstack endpoint list
+----------------------------------+-----------+--------------+----------------+---------+-----------+---------------------------------------------+
| ID                               | Region    | Service Name | Service Type   | Enabled | Interface | URL                                         |
+----------------------------------+-----------+--------------+----------------+---------+-----------+---------------------------------------------+
| 0081946f15cb48849331297444dd0442 | RegionOne | neutron      | network        | True    | admin     | http://controller:9696                      |
| 07bd808888764d24b50e15144e8a6fe8 | RegionOne | ceilometer   | metering       | True    | internal  | http://controller:8777                      |
| 182a6f4f3153441c876164c8dd0d9c3b | RegionOne | neutron      | network        | True    | internal  | http://controller:9696                      |
| 258207871d6e456ea1bb7d62c3caf48d | RegionOne | keystone     | identity       | True    | admin     | http://controller:35357/v3                  |
| 25ce52947a7942e0a43564d3445811b1 | RegionOne | keystone     | identity       | True    | public    | http://controller:5000/v3                   |
| 2e2b0b279b894cfe8cc08921ddb5e654 | RegionOne | heat         | orchestration  | True    | public    | http://controller:8004/v1/%(tenant_id)s     |
| 30190b8009984fada23336af8d9b4d42 | RegionOne | glance       | image          | True    | public    | http://controller:9292                      |
| 3768e82d1cdb413fbc51878c851dc70a | RegionOne | neutron      | network        | True    | public    | http://controller:9696                      |
| 37b6e7c2de7146c3ab828a8216bccc30 | RegionOne | glance       | image          | True    | admin     | http://controller:9292                      |
| 3d3fb15f11a84bfbb3c7b3267036547b | RegionOne | heat-cfn     | cloudformation | True    | internal  | http://controller:8000/v1                   |
| 3da58f94618844e8b50491ae0a585a3b | RegionOne | heat-cfn     | cloudformation | True    | admin     | http://controller:8000/v1                   |
| 432b9c9a5b784e6f9387594bd5bc2fba | RegionOne | aodh         | alarming       | True    | internal  | http://controller:8042                      |
| 52b82bcc8fcb48458d8fdc26e345eb96 | RegionOne | cinder       | volume         | True    | internal  | http://controller:8776/v1/%(tenant_id)s     |
| 55d984e4d86e4b32b784b3aba31cb0b1 | RegionOne | aodh         | alarming       | True    | admin     | http://controller:8042                      |
| 56437f028b324b19bea164c23c1a0224 | RegionOne | heat         | orchestration  | True    | internal  | http://controller:8004/v1/%(tenant_id)s     |
| 58689480ad4b4332a3b608ca16e72134 | RegionOne | heat         | orchestration  | True    | admin     | http://controller:8004/v1/%(tenant_id)s     |
| 5e647be1753849569fe8ceee0a60751b | RegionOne | cinderv2     | volumev2       | True    | public    | http://controller:8776/v2/%(tenant_id)s     |
| 80a46b7dd8aa417b8eaa5bcc44bcaa69 | RegionOne | cinderv2     | volumev2       | True    | internal  | http://controller:8776/v2/%(tenant_id)s     |
| 879a7b30673c4ae99ea33f3968ebc3ed | RegionOne | swift        | object-store   | True    | internal  | http://controller:8080/v1/AUTH_%(tenant_id)s |
| 8a4dd6ab28f24f04a252db7731e191d4 | RegionOne | nova         | compute        | True    | internal  | http://controller:8774/v2.1/%(tenant_id)s   |
| 8bbd9673162f415c96e65fc75e1cdebb | RegionOne | ceilometer   | metering       | True    | public    | http://controller:8777                      |
| 908ed9af43094b2f9769abe489a4f2d2 | RegionOne | keystone     | identity       | True    | internal  | http://controller:5000/v3                   |
| 94336af80e704899955f496bc1499e6b | RegionOne | cinder       | volume         | True    | public    | http://controller:8776/v1/%(tenant_id)s     |
| 9a998f6520294d328db13f91fb9fea78 | RegionOne | aodh         | alarming       | True    | public    | http://controller:8042                      |
| 9b67f826eccb4a2fa99493d529bacc6a | RegionOne | nova         | compute        | True    | public    | http://controller:8774/v2.1/%(tenant_id)s   |
| 9d6c103c916c4dfd90b1f36d7643475b | RegionOne | cinderv2     | volumev2       | True    | admin     | http://controller:8776/v2/%(tenant_id)s     |
| c8639492a3144feba9af1fa024fb57f8 | RegionOne | swift        | object-store   | True    | admin     | http://controller:8080/v1                   |
| c91a6411cde14194bf14a640e7faadc5 | RegionOne | swift        | object-store   | True    | public    | http://controller:8080/v1/AUTH_%(tenant_id)s |
| caf54d7724cb4bf48a14f719a4bfe6ff | RegionOne | heat-cfn     | cloudformation | True    | public    | http://controller:8000/v1                   |
| d6f29007f85942b49a3d593cf401aba7 | RegionOne | ceilometer   | metering       | True    | admin     | http://controller:8777                      |
| d7ce3efc55df4714b08bd180bd700985 | RegionOne | cinder       | volume         | True    | admin     | http://controller:8776/v1/%(tenant_id)s     |
| f05f8cf412dd47b5a5260f955217f982 | RegionOne | glance       | image          | True    | internal  | http://controller:9292                      |
| fb45df3033374effa7dd3399e8b7bc01 | RegionOne | nova         | compute        | True    | admin     | http://controller:8774/v2.1/%(tenant_id)s   |
+----------------------------------+-----------+--------------+----------------+---------+-----------+---------------------------------------------+
[root@controller ~]#
```

图 7.4 查看 cinder-api 的端点列表信息

查看 cinderv2 的端点信息，执行如下命令。

```
# openstack  endpoint  list  --service  cinderv2
```

其命令的结果如图 7.5 所示。

```
[root@controller ~]# openstack endpoint list --service cinderv2
| ID                               | Region    | Service Name | Service Type | Enabled | Interface | URL                                      |
| 5e647be1753849569fe8ceee0a60751b | RegionOne | cinderv2     | volumev2     | True    | public    | http://controller:8776/v2/%(tenant_id)s  |
| 80a46b7dd8aa417b8eaa5bcc44bcaa69 | RegionOne | cinderv2     | volumev2     | True    | internal  | http://controller:8776/v2/%(tenant_id)s  |
| 9d6c103c916c4dfd90b1f36d7643475b | RegionOne | cinderv2     | volumev2     | True    | admin     | http://controller:8776/v2/%(tenant_id)s  |
[root@controller ~]#
```

图 7.5 查看 cinderv2 的端点信息

客户端可以将请求发送到端点指定的地址，向 cinder-api 请求卷的操作。当然，用户不会直接发送 REST API 请求，而是由 OpenStack 命令行、Dashboard 以及其他需要与 Cinder 交互的 OpenStack 组件来使用这些 API 请求。

cinder-api 提供了 REST 标准调用服务，以方便与第三方系统集成。通过运行多个 cinder-api 进程可实现 API 的高可用性。

5. cinder-scheduler 服务

Cinder 可以有多个存储节点，当需要创建卷时，cinder-scheduler 将请求转到合适的 cinder-volume 服务，通过调度算法选择最合适的存储节点以创建卷。根据配置，可以通过简单的轮询调度，也可以通过过滤调度器实现更复杂的调度。默认使用过滤调度器，可以基于容量（Capacity）、可用区域（Availability Zone）、卷类型（Volume Type）和计算能力（Capabilities）进行过滤，也可以自定义过滤器。

cinder-scheduler 与 Nova 中的 nova-scheduler 的运行机制完全一样。其先通过过滤器选择满足条件的存储节点运行 cinder-volume，再通过权重计算选择最优的存储节点。可以在 Cinder 的配置文件/etc/cinder/cinder.conf 中对 cinder-scheduler 进行配置。

（1）默认的调度器 FilterScheduler

目前 Cinder 只实现了一个调度器 FilterScheduler，这也是 cinder-scheduler 默认的调度器，/etc/cinder/cinder.conf 文件中的默认配置如下。

scheduler_driver=cinder.scheduler.filter_scheduler.FilterScheduler

与 Nova 一样，Cinder 也允许使用第三方调度器，只需要配置 scheduler_driver 即可，需要注意的是，不同的调度器不能共享。

（2）过滤器设置

当调度器 FilterScheduler 需要执行调度操作时，会让过滤器对存储节点进行判断，满足条件返回 True，否则返回 False。在/etc/cinder/cinder.conf 文件中使用 scheduler_default_filters 参数指定使用的过滤器，默认配置如下。

scheduler_default_filters= AvailabilityZoneFilter, CapacityFilter, CapabilitiesFilter

调度器 FilterScheduler 将按照列表中的顺序依次进行过滤。

① 可用区域过滤器（Availability Zone Filter）。为提高容灾能力和提供隔离服务，可以将存储节点划分到不同的可用区域中。OpenStack 默认有一个命名为 Nova 的可用区域，所有的节点都默认放在 Nova 区域中，用户可以根据需要创建自己的可用区域。

② 容量过滤器（Capacity Filter）。创建卷时用户会指定卷的大小，CapacityFilter 的作用是将存储空间不能满足卷创建需要的存储节点过滤掉。

③ 能力过滤器（CapabilitiesFilter）。基于实例和卷资源类型记录的后端过滤器，不同的卷提供者有自己的能力，如是否支持精简置备。Cinder 允许用户创建卷时通过卷类型来指定所需要的能力。卷类型可以根据需要定义若干能力来详细描述卷的属性，卷类型的作用与 Nova 的实例类型（Flavor）类似。

（3）权重计算

经过调度器 FilterScheduler 的过滤，FilterScheduler 选出了能够创建卷的存储节点。如果有多个存储节点通过了过滤，那么最终选择哪个节点还需要进一步的确定。对这些节点计算权重值并进行排序，得出一个最佳的存储节点。这个过程需要调用权重计算模块，在/etc/cinder/cinder.conf 文件中通过 scheduler_default_weighers 参数指定权重过滤器，默认为 CapacityWeigher，默认配置如下。

```
scheduler_default_weighers=CapacityWeigher
```

CapacityWeigher 基于存储节点的空闲容量计算权重值，空闲容量最大的会被选中。

6. cinder-volume 服务

调度服务只负责分配任务，而真正执行任务的是 Worker 服务。cinder-volume 就是 Cinder 的 Worker。这种 Scheduler 和 Worker 之间功能上的划分使 OpenStack 易于扩展。一方面，当存储资源不够时，可以增加存储节点，即增加 Worker；另一方面，当客户端的请求量太大而又调度不过来时，可以增加 Scheduler 部署。

cinder-volume 在存储节点上运行，OpenStack 对卷的生命周期的管理最后都会交给 cinder-volume 完成，包括卷的创建、扩展、连接、快照、删除等操作。

cinder-volume 自身并不管理实际的存储设备，存储设备是由卷驱动（Volume Drivers）管理的，cinder-volume 与卷驱动一起实现卷的生命周期管理。在 Cinder 的驱动架构中，运行 cinder-volume 的存储节点和卷提供者可以是完全独立的两个实例。cinder-volume 通过驱动与卷提供者通信，控制管理卷。

（1）通过卷驱动架构支持多种后端存储设备

为支持不同的后端存储技术和设备，Cinder 提供了一个驱动框架，为这些存储设备定义了统一接口，如图 7.6 所示，第三方存储设备只需要实现这些接口，就可以以驱动的形式加入 OpenStack。

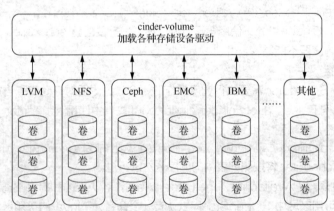

图 7.6 存储设备驱动架构

目前 Cinder 支持多种后端存储设备，包括 LVM、NFS、Ceph、EMC、IBM 等商业存储系统。

可以在/etc/cinder/cinder.conf 文件中使用 volume_driver 参数指定使用哪种后端存储设备，默认配置如下。

```
volume_driver=cinder.volume.drivers.lvm.LVMVolumeDriver
```

这表示 Cinder 默认使用本地的 LVM 逻辑卷。LVM 逻辑卷采用的是一种基于物理驱动器创建逻辑驱动器的机制，主要用于弹性地调整文件系统的容量，可以实现动态分区。

（2）多存储后端

Cinder 可以同时支持多个或多种后端存储设备，为同一个计算服务提供存储服务。Cinder 为每一个后端或后端存储池运行一个 cinder-volume 服务。

在多存储后端配置中，每个后端都有一个名称。多个后端可能会用一个名称，在这种情况下，由调度服务决定选用哪个后端来创建卷。后端的名称是作为卷类型的一个扩展规格来定义的，如"volume_backen_name=LVM"。创建卷时，调度服务会根据用户选择的卷类型选择一个合适的后端来处理请求。

要使用多存储后端，必须在/etc/cinder/cinder.conf 配置文件中使用 enabled_backends 参数定义不同后端的配置名称，多个名称由逗号分隔，其中一个名称关联一个后端的配置组。注意，配置组名称与卷后端名称没有关系。

对一个已知的 Cinder 服务设置 enabled_backends 参数后，重启块存储服务，则原来的主机服务将被新的主机服务替换，新的服务将以 host@backend 形式的名称出现，如 controller@lvm。

一个配置组的选项或参数必须在该组中定义，所有标准块存储配置参数都可以在配置组中使用。这样在配置组中的配置将被特定配置组中相同参数值所替换。需要注意的是，不同类型的后端需要定义不同的配置组参数。

（3）卷类型

Cinder 的卷类型的作用与 Nova 的实例类型（Flavor）类似，存储后端的名称需要通过卷类型的扩展规格来定义。创建一个卷后，必须指定卷类型，因为卷类型的扩展规格决定了要使用的后端。

使用卷类型之前必须先定义，再创建一个扩展规格，将卷类型连接到后端名称。需要注意的是，如果一个卷类型指向一个块存储配置中不存在的卷后端名称，那么过滤调度器将返回一个错误，提示不能找到具有合适后端的有效主机。

可以根据需要创建卷类型，并通过创建扩展规格来为该类型指定卷后端名称。

（4）卷连接到虚拟机

存储节点和计算节点往往是不同的物理节点，位于存储节点的卷与位于计算节点的虚拟机实例之间一般通过 ISCSI 协议进行连接。

在 OpenStack 中，cinder-volume 创建的卷可以以 ISCSI 目标的方式提供给 Nova。向 cinder-api 发送连接请求，cinder-volume 服务通过 ISCSI 协议将该卷连接到计算节点上，供虚拟机实例使用，如图 7.7 所示。

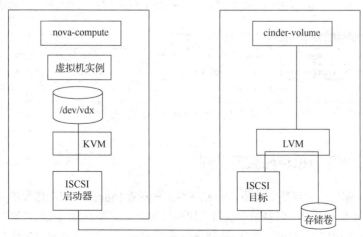

图 7.7　卷连接到虚拟机

Cinder 支持多种提供 ISCSI 目标的方法，可以在/etc/cinder/cinder.conf 文件中使用 iscsi_helper 参数进行配置。

连接卷的工作流程如下。

① 向 cinder-api 发送连接请求。

② cinder-api 发送消息。

③ cinder-volume 初始化卷的连接。

④ nova-compute 将卷连接到实例。

（5）cinder-volume 定期报告存储节点状态

cinder-scheduler 会用 CapacityFilter 和 CapacityWeigher 基于剩余容量来过滤存储节点，存储节点的空闲容量信息由 cinder-volume 提供，cinder-volume 会定期向 Cinder 报告当前存储节点的资源使用情况。

7. cinder-backup 服务

cinder-backup 为卷提供备份和恢复功能，实现了基于块的容灾。从 OpenStack 的 Kilo 版本开始，Cinder 引入了增量备份功能，相对全量备份复制和传输整个数据卷，增量备份只需要传输变化的部分，大大节省了传输开销和存储开销。

cinder-backup 支持块存储卷备份到 OpenStack 对象存储，目前支持的备份存储系统有 Ceph、NFS、POSIX、Swift 等。与 cinder-volume 通过卷驱动架构支持多种后端存储设备类似，cinder-backup 使用备份驱动架构来支持这几种备份存储系统。它通过 cinder.conf 配置文件中的 backup_driver 参数来指定使用的备份驱动，默认设置如下。

```
backup_driver=cinder.backup.drivers.swift
```

这表明默认将卷备份到 Swift 存储系统中。

8. Cinder 的物理部署

Cinder 的组件或子服务可以部署在控制节点和存储节点上。cinder-api 和 cinder-scheduler 部署在控制节点上，而 cinder-volume 部署在存储节点上。相关的 RabbitMQ 和 MySQL 通常部署在控制节点上，也可以将所有的 Cinder 服务都部署在同一节点上。在生产环境中，通常要将 OpenStack 服务部署在多台物理机上，以获得更好的性能和高可用性。

卷提供者是独立部署的。cinder-volume 使用驱动与它通信并协调工作，所以通常将驱动与 cinder-volume 放到一起。

使用命令 cinder service-list 可以查看 Cinder 子服务分布在哪些节点上，如图 7.8 所示。

```
[root@controller ~]# cinder service-list
+------------------+-----------------+------+---------+-------+----------------------------+-----------------+
|      Binary      |       Host      | Zone |  Status | State |         updated_at         | Disabled Reason |
+------------------+-----------------+------+---------+-------+----------------------------+-----------------+
| cinder-scheduler |    controller   | nova | enabled |   up  | 2021-01-21T06:54:05.000000 |        -        |
|  cinder-volume   | controller@lvm  | nova | enabled |   up  | 2021-01-21T06:54:03.000000 |        -        |
+------------------+-----------------+------+---------+-------+----------------------------+-----------------+
[root@controller ~]#
```

图 7.8 查看 Cinder 子服务分布

7.2.2　Swift 对象存储服务基础

Swift 对象存储是一个系统，可以上传和下载，一般存储的是不经常修改的内容，例如，存储虚拟机镜像、备份和归档，以及其他文件（如照片和电子邮件消息），它更倾向于系统的管理。Swift 可以将对象（可以理解为文件）存储到命名空间 Bucket（可以理解为文件夹）中，用 Swift

创建容器 Container，然后上传文件，如视频、照片等，这些文件会被复制到不同的服务器中，以保证其可靠性，Swift 可以不依靠虚拟机工作。所谓云存储，在 OpenStack 中就是通过 Swift 来实现的，可以把它理解成一个文件系统。Swift 作为一个文件系统，意味着可以为 Glance 提供存储服务，同时可以为个人的网盘应用提供存储支持，这个优势是 Cinder 和 Glance 无法实现的。

Swift 对象存储提供高可用性、分布式、最终一致性的对象存储，可高效、安全和廉价地存储大量数据。

1. Swift 对象存储

Swift 对象存储适合存储静态数据，所谓静态数据，是指长期不会发生变化更新，或者在一定时间内更新频率较低的数据，在云中主要有虚拟机镜像、多媒体数据，以及数据的备份。对于需要实时更新的数据，Cinder 块存储是更好的选择。Swift 通过使用标准化的服务器集群来存储 PB 数量级的数据，它是海量静态数据的长期存储系统，可以检索和更新这些数据。

Swift 对象存储使用了分布式架构，没有中央控制节点，可提供更高的可扩展性、冗余性等性能。对象写入多个硬件设备，OpenStack 负责保证集群中的数据复制和完整性。可通过添加新节点来扩展存储集群。当节点失效时，OpenStack 将从其他正常运行的节点复制内容。因为 OpenStack 使用软件逻辑来确保它在不同的设备之间的数据复制和分布，所以可以用廉价的硬盘和服务器来代替昂贵的存储设备。

对象存储是高性价比、可扩展存储的理想解决方案，提供一个完全分布式、API 可访问的平台，可以直接与应用集成，或者用于备份、存档和数据保存。Swift 适用于许多应用场景，最典型的应用是作为网盘类产品的存储引擎。在 OpenStack 中，其还可以与镜像服务 Glance 结合存储镜像文件。另外，由于 Swift 具有无限扩展能力，它也非常适合存储日志文件和数据备份仓库。

与文件系统不同，对象存储系统所存储的逻辑单元是对象，而不是传统的文件。对象包括了内容和元数据两个部分。与其他 OpenStack 项目一样，Swift 提供了 REST API 作为公共访问的入口，每个对象都是一个 RESTful 资源，拥有唯一的 URL，可以通过它请求对象，可以直接通过 Swift API，或者使用主流编程语言的函数库来操作对象存储，如图 7.9 所示，但对象最终会以二进制文件的形式保存在物理存储节点上。

V7-3 Swift 对象存储

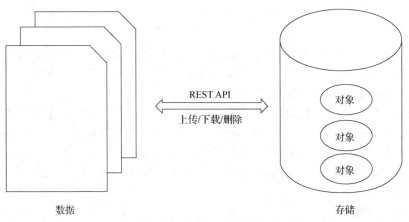

图 7.9 REST API 与存储系统交互

2. Swift 的系统架构

V7-4　Swift 的系统架构

Swift 采用完全对称、面向资源的分布式架构设计，所有组件均可扩展，避免因单点故障扩散而影响整个系统的运行。完全对称意味着 Swift 中各节点可以完全对等，能极大地降低系统维护成本。它的扩展性包括两个方面：一方面是数据存储容量无限可扩展；另一方面是 Swift 性能可线性提升，如吞吐量等。因为 Swift 是完全对称的架构，扩容只需简单地新增机器，所以系统会自动完成数据迁移等工作，使各存储节点重新达到平衡状态。Swift 的系统架构如图 7.10 所示。

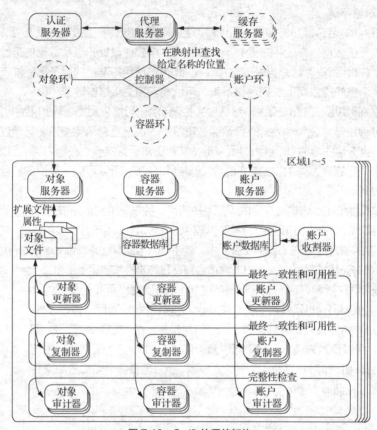

图 7.10　Swift 的系统架构

代理服务器为 Swift 其他组件提供统一的接口，它接收创建窗口、上传对象或者修改元数据的请求，还可以提供容器或者展示存储的文件。当收到请求时，代理服务器会确定账户、容器或者对象在容器环中的位置，并将请求转发到相关的服务器。

对象服务器上传、修改或检索存储在它所管理的设备上的对象（通常为文件）。容器服务器会处理特定容器的对象分配，并根据请求提供容器列表，还可以跨集群复制该列表。账户服务器通过使用对象存储服务来管理账户，其操作类似容器服务器的操作。

更新、复制和审记内部管理流程用于管理数据存储，其中复制服务最为关键，用于确保整个集群的一致性和可用性。

3. Swift 的应用

（1）网盘

Swift 的对称分布架构和多 Proxy 多节点的设计使它适用于多用户并发的应用模式，最典型的莫过于网盘的应用。

Swift 的对称架构使数据节点从逻辑上看处于同一级别，每个节点上同时具有数据和相关的元数据，并且元数据的核心数据结构使用的是哈希环，对于节点的增减一致性哈希算法只需要重定位环空间中的一小部分数据，具有较好的容错性和可扩展性。另外，数据是无状态的，每个数据在磁盘上都是完整的，这些特点保证了存储本身的良好的扩展性。在与应用的结合上，Swift 是遵循 HTTP 的，这使应用和存储的交互变得简单，不需要考虑底层基础构架的细节，应用软件不需要进行任何的修改就可以让系统整体扩展到非常大的程度。

（2）备份文档

Rackspace 是全球三大云计算中心之一，是一家成立于 1998 年的全球领先的托管服务器及云计算提供商，公司总部位于美国，它在英国、澳大利亚、瑞士、荷兰等地设有分部，在全球拥有 10 个以上的数据中心，管理超过 10 万台服务器，为全球 150 多个国家的客户服务。Rackspace 的托管服务产品包括专用服务器、云服务器、电子邮件、SharePoint、云存储、云网站等。Rackspace 的主营业务就是数据的备份归档，同时，其延展出一种新业务，如"热归档"。由于长尾效应，数据可能被调用的时间越来越长，"热归档"能够保证应用归档数据在分钟级别重新获取，和传统磁带机归档方案中的数小时相比是一个很大的进步。长尾效应中的"头"（Head）和"尾"（Tail）是两个统计学名词，正态曲线中间的突起部分叫"头"，两边相对平缓的部分叫"尾"。从人们需求的角度来看，大多数的需求会集中在头部，而这部分可以被称为"流行"，分布在尾部的需求是个性化的、零散的、小量的需求。这部分差异化的、少量的需求会在需求曲线上面形成一条长长的"尾巴"，而所谓长尾效应就在于它的数量，将所有非流行的市场累加起来就会形成一个比流行市场还大的市场。

（3）IaaS 公有云

Swift 在设计中的线性扩展、高并发和多项目支持等特性，使它非常适合作为 IaaS。公有云规模较大，更多时候会遇到大量虚拟机并发启动的情况，所以对于虚拟机镜像的后台存储来说，实际上的挑战在于大数据的并发读性能。Swift 在 OpenStack 中一开始就是作为镜像库的后台存储，经过 Rackspace 上千台机器的部署规模下的数年实践，Swift 已经被证明是一个成熟的选择方案。另外，基于 IaaS 要提供上层的 SaaS，多项目是一个不可避免的问题，Swift 的架构设计本身就是支持多项目的，这样对接起来更方便。

（4）移动互联网络和内容分发网络

移动互联网和手机游戏等会产生大量用户数据，单个用户的数据量虽然不是很大，但是用户数很多，这也是 Swift 能够处理的领域。

内容分发网络（Content Delivery Network，CDN）是构建在现有网络基础之上的智能虚拟网络，依靠部署在各地的边缘服务器，通过中心平台的负载均衡、内容分发、调度等功能模块，使用户就近获取所需内容，降低网络拥塞，提高用户访问响应速度和命中率。CDN 的关键技术是内容存储和分发。

至于 CDN，如果使用 Swift，云存储就可以直接响应移动设备，不需要专门的服务器去响应这个 HTTP 请求，也不需要在数据传输中再经过移动设备上的文件系统，而是直接通过 HTTP 上传云端。如果把经常被平台访问的数据缓存起来，利用一定的优化机制，数据可以从不同的地点分发到用户那里，这样就能提高访问的速度。在 Swift 的开发社区中有人讨论视频网站应用和 Swift 的结合，这也是值得关注的方向。

4. Swift 的层次数据模型

Swift 将抽象的对象与实际的具体文件联系起来，需要使用一定方法来描述对象。Swift 采用的是层次数据模型，如图 7.11 所示。存储的对象在逻辑上分为 3 个层次——账户、容器和对象，每一层所包含的节点数没有限制，可以任意扩展。

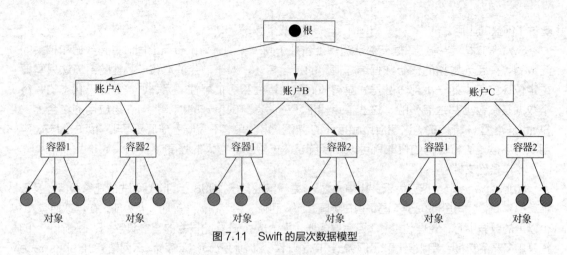

图 7.11　Swift 的层次数据模型

（1）账户

账户在对象存储过程中用于实现顶层的隔离，它并非个人账户，而指项目（或租户），可以被多个用户账户共同使用。账户由服务提供者创建，用户在账户中拥有全部资源，账户为容器定义了一个命名空间。Swift 要求对象必须位于容器中，因此一个账户应当至少拥有一个容器来存储对象。

（2）容器

容器表示封装的一组对象，与文件夹或目录类似，不过容器不能嵌套，不能再包含下一级容器。容器为对象定义了命名空间，两个不同的容器中同一名称的对象代表两个不同的对象。除了包含对象外，也可以通过访问控制列表使用容器来控制对象的访问。在容器层级还可以配置和控制许多其他特性，如对象版本。

（3）对象

对象位于最底层，是叶子节点，具体的对象由元数据和内容两部分组成。对象存储诸如文档、图像这样的数据内容，可以为一个对象保存定制的元数据。Swift 对于单个上传对象有大小的限制，默认是 5GB。不过由于使用了分割的概念，单个对象的下载大小几乎没有限制。

（4）对象层级结构与对象存储 API 的交互

账户、容器和对象层级结构影响与对象存储 API 交互的方式，尤其是资源路径的层次结构。资源路径具有以下格式。

/v1/{account}/{container}/{object}

例如，对于账户 user01 的容器 images 中的对象 photo/flower.jpg，资源路径如下。

/v1/user01/images/photo/flower.jpg

对象名包含字符"/"，该字符并不表示对象存储有一个子级结构，因为容器不存储在实际子文件夹的对象中，但是在对象名中包括类似的字符可以创建一个伪层级的文件夹和目录。例如，如果对象存储为 object.mycloud.com，则返回的 URL 是 https:// object.mycloud.com/v1/user01。

要访问容器，应将容器名添加到资源路径中；要访问对象，应将容器名和对象名添加到资源路径中。如果有大量的容器或对象，则可以使用查询参数对容器或对象的列表进行分页。使用 maker、limit 和 end_marker 查询参数来控制要返回的条目数以及列表起始处。

/v1/{account}/{container}/?marker=a&end_marker=d

如果需要逆序，则可使用查询参数 reverse，注意 marker 和 end_marker 应当交换位置，以返回一个逆序列表。

/v1/{account}/{container}/?marker=d&end_marker=a&reverse=on

5. Swift 的组件

Swift 使用代理服务器（Proxy Servers）、环（Rings）、区域（Zones）、账户（Account）、容器（Container）、分区（Partitions）等组件来实现高可用性、高持久性和高并发性。Swift 的部分组件如图 7.12 所示。

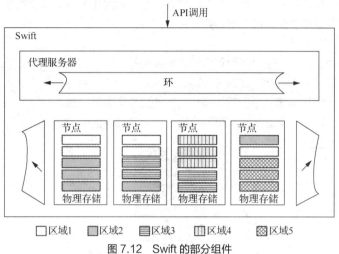

图 7.12　Swift 的部分组件

（1）代理服务器

代理服务器是对象存储的公共接口，用于处理所有传入的请求。一旦代理服务器收到一个请求，它就会根据对象 URL 决定存储节点。代理服务器也负责协调响应、处理故障和标记时间戳（Timestamp）。

代理服务器使用无共享架构，能够根据预期的负载按需扩展。在一个单独管理的负载平衡集群中应最少部署两台代理服务器，如果其中一台出现故障，则由其他代理服务接管。

（2）环

环表示集群中保存的实体名称与磁盘上物理位置之间的映射，将数据的逻辑名称映射到特定磁盘的具体位置。账户、容器和对象都有各自的环。当系统组件需要对象、容器或账户执行任何操作时，都需要与相应的环进行交互，以确定其在集群中的合适位置。

环使用区域（Zone）、设备（Device）、分区（Partition）和副本（Replicas）来维护这种映射信息。每个分区在环中都有副本，默认集群中有 3 个副本，存储在映射中的分区的位置由环来维护，如图 7.13 所示。环也负责决定在发生故障时使用哪个设备接收请求。

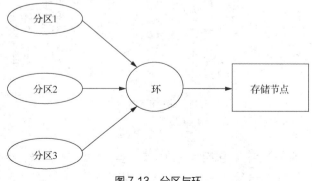

图 7.13　分区与环

环中的数据被隔离到区域中。每个分区的副本设备都存储在不同的区域中，区域可以是一个驱动器、一台服务器、一个机柜、一台交换机，甚至是一个数据中心。在对象存储安装过程中，环的分区会均衡地分配到所有的设备中。当分区需要移动时，如新设备被加入集群，环会确保一次移动最少数量的分区数，并且一次只移动一个分区的一个副本。

权重可以用来平衡集群中驱动器的分布。例如，当不同大小的驱动器被用于集群时就显得非常必要。环由代理服务器和一些后台使用，如复制进程等。

（3）区域

为隔离故障边界，对象存储允许配置区域。如果可能，每个数据副本位于一个独立的区域。最小级别的区域可以是一个驱动器或一组驱动器。如果有 5 个对象存储服务器，则每个服务器将代表自己的区域。可以大规模部署一个机架或多个机架的对象服务器，每个对象服务器代表一个区域。因为数据跨区复制，所以一个区域的故障不影响集群中的其余区域，如图 7.14 所示。

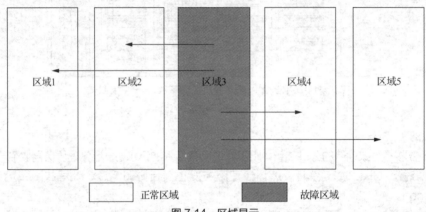

图 7.14　区域显示

（4）账户和容器

每个账户和容器都是独立的 SQLite 数据库，这些数据库在集群中采用分布式部署。账户数据库包括该账户中的容器列表，容器数据库包含该容器中的对象列表。账户和容器的关系如图 7.15 所示。为跟踪对象数据位置，系统中的每个账户都有一个数据库，它引用其全部容器，每个容器数据库可以引用多个对象。

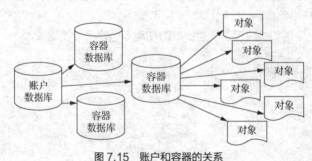

图 7.15　账户和容器的关系

（5）分区

分区是存储的数据的一个集合，包括账户数据库、容器数据库和对象，有助于管理数据在集群中的位置，如图 7.16 所示。对于复制系统来说，分区是核心。

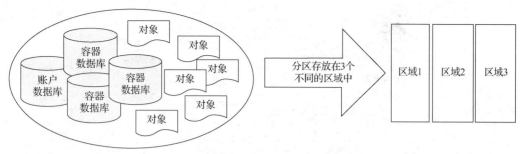

图 7.16　分区

可以将分区看作在整个中心仓库中移动的箱子。个别的"订单"投进箱子,系统将该箱子作为一个紧密结合的整体在系统中移动。这个箱子比许多小物件更容易处理,有利于在整个系统中较少地移动。

系统复制和对象下载都是在分区上操作的,当系统扩展时,其行为是可以预测的,因为分区数量是固定的。分区的实现概念很简单,一个分区就是位于磁盘上的一个目录,拥有它所包含内容的相应哈希表。

(6)复制器

为确保始终存在 3 个数据副本,复制器(Replicators)会持续检查每个分区。对于每个本地分区,复制器将它与其他区域中的副本进行比较,确认是否发生变化,如图 7.17 所示。

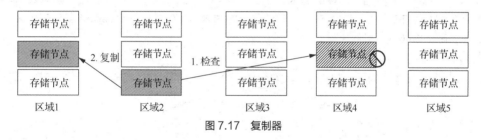

图 7.17　复制器

复制器通过检查哈希表来确认是否需要进行复制。每个分区都会产生一个哈希文件,该文件包含该分区中每个目录的哈希值。对于一个给定的分区,它的每个副本的哈希文件都会进行比较。哈希值不同,则需要复制,而需要复制的文件的目录也要进行复制。

这就是分区的用处,在系统中通过较少的工作传输更多的数据块,一致性比较强。集群具有最终一致性行为,旧的数据由错过更新的分区提供,但是复制会导致所有的分区向最新数据聚集。

如果一个分区出现故障,则一个包含副本的节点会发出通知,并主动将数据复制到接管的节点上。

(7)对象存储组件的协同工作

对象存储中的对象上传和下载如图 7.18 所示。从图中可知相关组件是如何协同工作的。

① 对象上传。

首先,客户使用 REST API 构造一个 HTTP 请求,将一个对象上传到一个已有的容器中,集群收到请求,系统必须解决将该对象存放到哪里的问题,通过负载平衡器的各代理节点进行计算。为此,账户名称、容器名称和对象名称都用来决定该对象的存放位置。

其次,通过环中的查询明确使用哪个存储节点来容纳该分区。

最后,数据被发送到要存放该分区的每个存储节点上。在客户收到上传成功通知之前,必须至少有三分之二的写入是成功的。接下来容器数据库异步更新,反映已加入的新对象的情况。

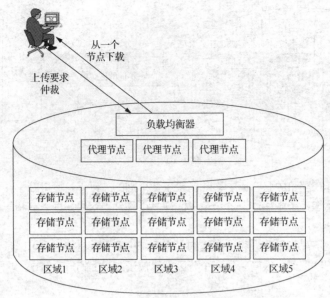

图 7.18 对象存储中的对象上传和下载

② 对象下载。

收到一个对账户、容器或对象的请求时,使用同样的一致性哈希算法来决定分区的索引。环中的查询获知哪个存储节点包含在该分区中。此后,将请求提交给其中一个存储节点来获取该对象,如果失败,则请求转发给其他节点。

6. 对象存储集群的层次架构

Swift 对象存储集群的层次架构可以大致分为两个——访问层和存储节点,如图 7.19 所示。

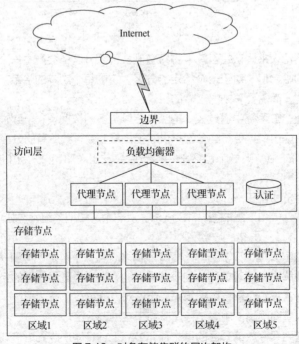

图 7.19 对象存储集群的层次架构

（1）访问层

大规模部署需要划分出一个访问层，将其作为对象存储系统的中央控制器。访问层接收客户传入的 API 请求，管理系统中数据的进出。该层包括前端负载均衡器、安全套接层（Secure Sockets Layer，SSL）终结前端和认证服务。它运行存储系统的中枢，即代理服务器进程。

由于访问服务器集群在自己所在的层，因此可以扩展读写访问，而与存储容量无关。例如，如果一个集群位于 Internet，要求 SSL 终结前端，而且对数据访问要求高，那么可以置备许多访问服务器。如果集群位于内部网络且主要用于存档，则只需要置备少量访问服务器。

既然这是一个 HTTP 可访问的存储服务，便可以将负载均衡器并入访问层。典型的访问层包括服务器的集合。这些服务器使用适量的内存，网络 I/O 能力很强。系统接收每个传入的 API 请求，应当配置两个高带宽的接口，一个用于接入的前端请求，另一个用于对象存储节点的后端访问以及提交或获取数据。

对于大多数面向公共的部署和大的企业网络的私有部署来说，必须使用 SSL 来加密客户端的流量。通过 SSL 建立客户之间的会话会大大增加处理负载。这就是不得不在访问层置备更多服务器的原因，在可信网络中没有必要部署 SSL。

（2）存储节点

在多数配置中，5 个区域中的每个区域都应当有相同的存储容量，存储节点使用合适的内存和 CPU，需要方便地获取元数据以快速返回对象。对象存储运行服务，不仅接收来自访问层的传入请求，还运行复制器、审计器和收割器。置备存储节点可以使用单个 1GB/s 或 10GB/s 的网络接口，这取决于预期的负载和性能。

目前，一块 3TB 或 5TB 的 SATA 硬盘具有很高的性价比。如果数据中心有远程操作，则可以使用桌面级驱动器，否则使用企业级驱动器。

应当考虑单线程请求所希望的 I/O 性能，此系统不用磁盘阵列，所以使用单块硬盘处理一个对象的每个请求，硬盘性能影响单线程响应速度。

要显著获得较大流量，对象存储系统应设计能处理并发的上传和下载，网络 I/O 能力（10GbE）应当满足读写所需的并发流量。

7. Swift 服务的优势

（1）数据访问灵活性

Swift 通过 REST API 来访问数据，可以通过 API 实现文件的存储和管理，使资源管理实现自动化。同时，Swift 将数据放置于容器内，用户可以创建公有的容器和私有的容器。自由的访问控制权限既允许用户间共享数据，也可以保存隐私数据。Swift 对所需要的硬件没有刻意的要求，可以充分利用商用的硬件节约单位存储的成本。

V7-5 Swift 服务的优势

（2）数据的持久可靠性

Swift 提供多重备份机制，拥有极高的数据可靠性，数据存放在高分布式的 Swift 中，几乎不会丢失。

（3）极高的可拓展性

Swift 通过独立节点来形成存储系统，它在数据量的存储上做到了无限拓展。另外，Swift 的性能也可以通过增加 Swift 集群来实现线性提升，所以 Swift 很难达到性能瓶颈。

（4）无单点故障

由于 Swift 的节点具有独立的特点，在实际工作时不会发生传统存储系统的单点故障。传统存储系统即使通过高可用性（High Availability，HA）来实现热备，在主节点出现问题时，还是会影响整个存储系统的性能。而在 Swift 中，数据的元数据是通过环算法随机均匀分布的，且元数据也会保存多份，对于整个 Swift 集群而言，没有单点的角色存在。

8. Swift 的代理服务器

与 Cinder 块存储系统不同，Swift 对象存储不能由普通用户直接操作。只能由云管理员在后端进行配置管理操作，涉及配置文件和命令行操作。这里结合前面描述的 Swift 系统架构，在进一步分析相应服务和组件的基础上，讲解其配置与管理。

代理服务涉及代理服务器、认证服务器和缓存服务器。

（1）代理服务器

代理服务器负责将 Swift 架构的其余部分连接在一起，提供 Swift API 的服务进程，负责 Swift 其他组件之间的相互通信。

对每个请求，代理服务器将在环中查询账户、容器或对象的位置，并且相应地转发请求。对于前向错误纠正技术类型策略，代理服务器负责对对象数据进行编码和解码。对外公开的 API 也通过代理服务器对外提供。它提供的 REST API 符合标准的 HTTP 规范，这使开发者可以快捷地构建定制的客户端，与 Swift 进行交互。其采用了无状态的 REST 请求协议，因此可以进行横向扩展来均衡负载。

代理服务器也会处理大量的故障，例如，如果一个服务器对一个对象的上传操作不可用，则它会向环请求一台替代服务器，并进行转发。

对象以流的形式流进或流出对象服务器，它们直接通过代理服务器流进或流出用户的请求，代理服务器并不缓存它们。

代理服务器扮演了类似 nova-api 的角色，负责接收并转发客户的 HTTP 请求。针对 3 个层次的对象，可以分为针对账户、容器和对象的操作，表 7.1 所示为 Swift 所支持的基本操作。

表 7.1　Swift 所支持的基本操作

资源类型	资源路径	GET	PUT	POST	DELETE	HEAD
账户	/account/	获取容器列表	—	—	—	获取账户元数据
容器	/account/container	获取对象列表	创建容器	更新容器元数据	删除容器	获取容器元数据
对象	/account/container/object	获取对象内容和元数据	创建、更新或复制对象	更新对象元数据	删除对象	获取对象元数据

代理服务器通常部署在控制节点上，也可以部署在能够连接存储节点的其他任何节点上，其配置文件为/etc/swift/proxy-server.conf。

（2）认证服务器

Swift 通过代理服务器接收用户请求时，首先需要通过认证服务器对用户的身份进行验证，只有验证通过后，代理服务器才会处理用户请求并做出响应。

Swift 支持外部和内部两种认证方式，前者是指通过 Keystone 服务器进行认证，后者是指通过 Swift 的中间件进行认证。无论使用哪种方式，用户都要向认证服务器提交自己的凭证，认证系统返回一个令牌和 URL。令牌在一定时间内一直有效，可以缓存下来直到过期。在有效期内，用户在请求头部加上令牌来访问 Swift 服务。

（3）缓存服务器

缓存的内容包括对象服务令牌、账户和容器的信息，但不包括缓存对象本身的数据。缓存服务器可采用集群，Swift 会使用一致性哈希算法来分配缓存地址。

9. Swift 的存储服务器

存储服务器是基于磁盘设备提供的,分为 3 种类型,分别是账户服务器、容器服务器和对象服务器,它们与代理服务器一起构成 Swift 的 4 个主要服务器。一般存储服务器部署在专门的存储节点上,与对象层次数据模型对应,具体由以下 3 种服务器提供相应的服务。

(1)账户服务器

账户服务器提供与账户相关的服务,包括所含容器的列表和账户的元数据。账户信息存储在 SQLite 数据库中,其配置文件为/etc/swift/account-server.conf。

(2)容器服务器

容器服务器提供与容器相关的服务,包括所含对象的列表和容器的元数据。它管理的是从容器到对象的单一映射关系,并不知道对象存放在哪个容器中,只知道特定容器中存储了哪些对象。对象列表以 SQLite 数据库文件形式存储,可以跨集群复制。容器服务器也会跟踪统计对象总数和该容器的存储用量等统计信息,其配置文件为/etc/swift/container-server.conf。

(3)对象服务器

对象服务器提供对象的存取和元数据服务。它是一个非常简单的二进制对象存储服务器,可以存储、检索和删除存储在本地设备上的对象。对象以二进制文件的形式存储在文件系统上,而其元数据作为文件系统的扩展属性来存储。这就要求底层文件系统选择支持文件扩展属性的对象服务器。

每个对象的存储所用路径来自该对象名的哈希值和操作的时间戳。最后一次写入总是"胜出",从而确保对象版本最新。被删除的对象也被作为该文件的一个特殊版本,这可以确保被删除的文件能被正确复制,旧版本不再因为故障重新出现,其配置文件为/etc/swift/object-server.conf。

10. 存储策略管理

Swift 通过创建多个副本实现冗余来提高数据的持久性,但是这要牺牲更多的存储空间,必须尽可能减少占用存储空间,为此 Swift 采用纠删码(Erasure Code,EC)技术。它将数据分块,再对每一块加以编码以减少对空间的需求,而且可以在某块数据损坏的情况下根据其他块的数据将其恢复,实际上这是通过消耗更多的计算资源和网络带宽来降低对存储的消耗。存储策略就是要解决 EC 技术和副本的共存问题。

(1)存储策略的概念

存储策略提供了一种方法,让对象存储提供者区分 Swift 部署的服务层次、特性和行为。Swift 中配置的每个存储策略通过一个抽象名称提供给客户。系统中每个设备被分配给一个或多个存储策略。这通过使用多个对象环来实现,每个存储策略都有自己的对象环,对象环可能包括一个硬件子集以实现特定的区分。

Swift 对采用不同存储策略的对象采用不同的存储方式,例如,一个对象存储部署可能有两个存储策略,一个要求有 4 个副本,另一个要求有 2 个副本,显然,前者服务级别较高。可以在一个存储策略中包含固态硬盘,应用该策略的用户都能获得较高的存储效率,还可以使用 EC 技术来定义一个冷存储层(Cold Storage Tier)。

(2)配置存储策略

使用存储策略的关键问题是"如何确定一个对象的存储策略"。Swift 要求存储策略基于每个容器定义,每个容器在创建时可指定一个特定存储策略,在该容器生命周期内都有效。当然,多个容器也可以关联到同一个存储策略,这种关联必须在创建容器时确立,而且不可改变。

配置存储策略的步骤。

① 编辑/etc/swift/swift.conf 文件来定义新的策略。

② 创建相应策略的对象环文件。

③ 创建特定策略的代理服务器配置设置，这一步是可选的。

（3）定义存储策略

每个策略由/etc/swift/swift.conf 文件中的一个节定义，节名必须采用[storage-policy:<N>]这样的格式，其中<N>是策略索引。此外，还要遵守以下规则。

如果没有声明索引为 0 的策略且未定义其他策略，则 Swift 将创建一个索引为 0 的默认策略。策略索引必须为正整数，且必须唯一。

（4）创建环

定义新的策略后，必须创建一个新的环。存储策略和对象环之间的映射是通过索引来建立的，索引为 0 的策略对应的.builder 文件名为 object.builder；索引为 1 的策略对应的.builder 文件名为 object-1.builder；以此类推。

（5）使用策略

只有在容器的初始创建过程中才能使用存储策略。一旦创建的容器有特定策略，存储在其中的所有对象就都要遵从该策略。

11. 一致性服务及管理

在实际的云环境中，存储服务还要能够自动处理故障。Swift 通过多个副本确保不会因为软硬件故障导致数据丢失，并通过存储策略来降低多个副本带来的存储资源消耗，但是同一数据在不同副本之间还存在不一致的问题。为此，Swift 通过审计器（Auditors）、更新器（Updaters）和复制器（Replicators）3 个服务器程序来提供一致性服务，查找并解决由数据损坏和硬件故障引起的错误，从而解决数据的一致性问题。

（1）审计器

审计器负责数据的审计，在本地服务器上持续扫描磁盘，检测账户、容器和对象的完整性。如果发现数据损坏，该文件就会被隔离，然后由复制器从其他节点获取对应的完好的副本来替代损坏的文件。如果出现其他错误，如在任何一个容器服务器中都找不到所需要的对象列表，则将错误记录在日志中。

账户的审计器在账户服务器配置文件/etc/swift/account-server.conf 中的[account-auditor]节中设置；容器的审计器在账户服务器配置文件/etc/swift/container-server.conf 中的[container-auditor]节中设置；对象的审计器在账户服务器配置文件/etc/swift/object-server.conf 中的[object-auditor]节中设置。

（2）更新器

更新器运行更新服务，负责处理那些失败或容器更新操作。在发生故障或系统高负荷的情况下，容器或账户中的数据不会被立即更新执行。如果更新失败，则该次更新在本地文件系统上会被加入队列，然后由更新器继续处理这些失败的更新工作。由账户更新器和容器更新器分别负责账户列表和对象列表。例如，假设一个容器服务器处于负载状态，此时一个新的对象加入系统。代理服务器成功地响应客户的请求，这个对象就立即变为可读的。但是容器服务器并没有更新对象列表，这样此次更新将进入队列行等待延后的更新。因此，容器列表不可能立即包含这个新对象。

在实际使用中，窗口的大小和更新器的运行频率一致，因为代理服务器会将列表请求转送给第一个响应的容器服务器，但是可能不会被注意到。当然，处于负载状态的服务器未必再去响应后续的列表请求，其他两个副本中有一个可能会处理这些列表请求。

与审计器一样，账户、容器和对象的更新器分别在对应的账户、容器和对象服务器配置文件中的[account-updater]、[container-updater]和[object-updater]节中设置。

（3）复制器

复制器运行复制进程，负责检测各节点上数据及其副本是否一致，当发现不一致时会将过时的副本更新为最新版本，并且负责将标记为删除的数据从物理介质上删除。复制的设计目的是在面对像网络中断或驱动器故障等临时性错误情况时可以保持系统的一致性状态。

复制进程将本地数据与每个远程副本进行比较，以确保它们都包含最新的版本。对象复制使用一个哈希列表来快速地比较每个分区的分段，容器和账户的复制组合使用哈希值和共享的高级水印算法进行版本比较。

与审计器一样，账户、容器和对象的复制器分别在对应的账户、容器和对象服务器配置文件中的[account-replicator]、[container-replicator]和[object-replicator]节中设置。

12. 环的创建和管理

对象最终都要以文件的形式存储到存储节点上，但是 Swift 没有传统文件系统中的路径、目录或文件夹这样的概念，而是使用环的概念来解决对象与真正的物理存储位置的映射或关联。

环是 Swift 中非常核心的组件，决定着数据如何在存储集群中分布。账户数据库、容器数据库和对象数据库都有独立的环，但是每个环以相同方式工作。这些环都可以在外部管理，服务器进程本身不能修改环，只能获得由其他工具修改的新的环。

（1）环的实现原理

要解决某个对象存储在哪个节点的问题，最常规的做法是采用哈希算法。如果存储节点固定，普通的哈希算法即可满足要求。Swift 需要通过增减存储节点来实现可扩展性，节点数量会发生变化，此时所有哈希值都会改变，普通哈希算法无法满足要求。为此，Swift 引入了"环"这个概念，利用一致性哈希算法构建环，来解决海量对象在无限节点上存放的寻址问题。

为减少增减节点所带来的数据迁移，Swift 在对象和存储节点的映射之间增加了分区的概念，使对象到存储节点之间的映射变成了对象到分区再到存储节点的映射。分区的数量一旦确定，在整个运行过程中都不会改变，因此对象到分区的映射关系不会改变。增加或减少节点时，只需要通过改变分区到存储节点之间的映射关系，即可实现数据迁移。相对于实际存储数据的物理节点，分区相当于虚拟存储节点，因此也有人将 Swift 的术语"分区"（Partition）称为虚拟节点。

所有的分区（虚拟节点）平衡地构成一个平面的环状结构，Swift 采用一致性哈希算法，通过计算可以将对象均匀分布到每个分区中，分区的数量采用 2 的 n 次幂表示，便于进行高效的位移计算。由哈希函数和二进制位移操作实现对象对分区的映射。

（2）环的数据结构

环的数据结构由以下 3 个顶层域组成。

① 集群中设备的列表。

② 设备 ID 列表，表示分区到设备的指派。

③ 一个表示 MD5 哈希值位移操作的位数，即用于对分区进行哈希计算。

（3）环文件

Swift 存储中使用的环文件用于各个存储节点记录存储对象与实际物理位置的映射关系。客户对 Swift 存储数据进行操作时，均通过环文件来定位实际的物理位置。账户服务器、容器服务器和对象服务器都有各自的环文件。

构建环的过程中会生成.builder 文件和相应的.ring.gz 文件。这些文件默认位于/etc/swift 文件夹中，如图 7.20 所示。

在生成新的环文件之前，会将原来的.builder 文件和.ring.gz 文件备份到 backups 文件夹中，如图 7.21 所示。

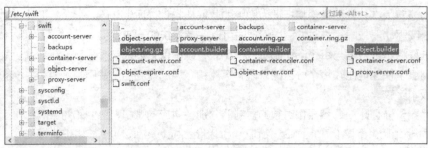

图 7.20 .builder 文件和 .ring.gz 文件

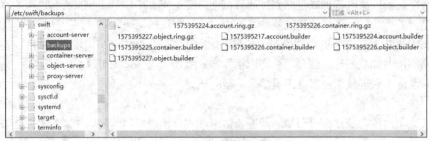

图 7.21 备份的 .builder 文件和 .ring.gz 文件

对 .builder 文件的保存非常重要，因此需要存储环文件的多个副本，因为一旦环文件完全丢失，就意味着需要完全从头创建环，这样几乎所有的分区都会被分配到新的不同的设备上，因此数据的副本也都会被移动到新的位置，造成大量的数据迁移，导致系统在一段时间内不可用。

（4）环的创建和管理

创建环是 Swift 初始化必须经历的过程。环是通过 swift-ring-builder 工具手动创建的，该工具将分区与设备关联，并将该数据写入一个优化过的 Python 数据结构，经压缩、序列化后写入磁盘，以使环创建的数据可以被导入服务器。更新环的机制非常简单，服务器通过检查创建环的文件的最后更新日期来判断它和内存中的版本哪一个是更新的，从而决定是否需要重新载入环创建的数据。

（5）分发环文件

创建环完成后需要向环中添加设备，通过重新分配最近没有被重新分配的分区来重新平衡环。在完成重新平衡环的操作后，需要将生成的环文件复制到所有运行存储服务的节点上，如账户服务器、容器服务器和对象服务器，并重新启动这些存储服务。

创建环的基本步骤总结如下。

① 创建环。
② 添加设备到环中。
③ 重新平衡环。
④ 分发环文件，并重启相关服务。

7.3 项目实施

7.3.1 基于 Web 界面管理存储服务

基于部署远程桌面管理 OpenStack 平台，可以验证和操作存储服务。用户以云管理员身份登录 Dashboard 界面，可以执行存储服务管理操作。

1. 卷管理

卷管理包括查看卷的列表信息、卷的创建、扩展云硬盘、创建云主机、管理连接、创建快照、修改云硬盘类型、上传镜像、创建转让、删除云硬盘、创建云硬盘类型、创建加密云硬盘类型、创建云硬盘类型扩展规格、创建 QoS 规格、QoS 规格关联到云硬盘类型等相关操作。

（1）查看项目中的卷列表信息

在 Dashboard 界面中，先单击"项目"主节点，再单击"计算"→"卷"子节点，打开卷列表界面，可以查看卷列表信息，如图 7.22 所示。

图 7.22　查看卷列表信息

（2）创建云硬盘

在卷列表界面右上方单击"+创建云硬盘"按钮，打开"创建云硬盘"对话框，云硬盘是可被连接到云主机的块设备。云硬盘类型描述为如果"无云硬盘类型"被选中，则云硬盘被创建的时候将没有云硬盘类型。如图 7.23 所示，创建一个名称为 cloud_volume-01 的云硬盘，云硬盘来源、类型等都不进行设置，默认进行创建即可。

图 7.23　创建云硬盘 cloud_volume-01

又如，创建一个名称为 cloud_volume-02 云硬盘，云硬盘来源设置为镜像，选择相应的镜像源，如图 7.24 所示。

图 7.24　创建云硬盘 cloud_volume-02

查看创建的云硬盘，可以看到名称为 cloud_volume-01 的云硬盘，其在"可启动"列下的状态为 False；而名称为 cloud_volume-02 的云硬盘在"可启动"列下的状态为 True，如图 7.25 所示。

图 7.25　创建的云硬盘 cloud_volume-01 与 cloud_volume-02 列表信息

（3）编辑云硬盘

选择相应的云硬盘，在卷列表界面的右侧的"操作"下拉列表中选择"编辑云硬盘"选项，打开"编辑云硬盘"对话框，如图 7.26 所示，可以修改云硬盘的名称和描述。"可启动"表明云硬盘可以被用来创建云主机。

图 7.26　编辑云硬盘

选择相应的云硬盘，在卷列表界面的右侧的"操作"下拉列表中，可以进行扩展云硬盘、创建云主机、管理连接、创建快照、修改云硬盘类型、上传镜像、创建转让、删除云硬盘等相关操作，如图 7.27 所示。

图 7.27　云硬盘相关操作

（4）扩展云硬盘

选择相应的云硬盘，在卷列表界面的右侧的"操作"下拉列表中选择"扩展云硬盘"选项，打开"扩展云硬盘"对话框，设置扩展云硬盘的大小，新大小必须大于当前设置，如图 7.28 所示。

图 7.28　扩展云硬盘

单击"扩展云硬盘"按钮,返回卷列表界面,可以看到 cloud_volume-02 的云硬盘磁盘容量由 2GiB 变为 3 GiB,如图 7.29 所示。

图 7.29　查看扩展云硬盘容量大小变化情况

（5）创建云主机

选择相应的云硬盘,在卷列表界面的右侧的"操作"下拉列表中选择"创建云主机"选项,打开"启动实例"对话框,设置实例名称、可用区域等,如图 7.30 所示。

图 7.30　创建云主机

（6）管理连接

选择相应的云硬盘,在卷列表界面的右侧的"操作"下拉列表中选择"管理连接"选项,打开"管理已连接云硬盘"对话框,选择要连接的云主机,如图 7.31 所示。

图 7.31　管理已连接云硬盘

单击"连接云硬盘"按钮，返回卷列表界面，可以看到云硬盘 cloud_volume-01 与 cloud_volume-02 已经连接到相应的云主机 cloud_host-01 与 cloud_host-10，如图 7.32 所示。

图 7.32　云硬盘连接相应云主机列表信息

（7）创建快照

选择相应的云硬盘，在卷列表界面的右侧的"操作"下拉列表中选择"创建快照"选项，打开"创建云硬盘快照"对话框，如图 7.33 所示。

图 7.33　创建云硬盘快照

输入快照名称等相关信息创建快照，单击"创建云硬盘快照（强制）"按钮，完成创建快照操作，返回"卷快照"选项卡，即可查看卷快照列表信息，如图 7.34 所示。

图 7.34　卷快照列表信息

选择相应的卷快照，在"卷快照"选项卡右侧的"操作"下拉列表中选择"删除云硬盘快照"选项，可以删除云硬盘快照，如图 7.35 所示。

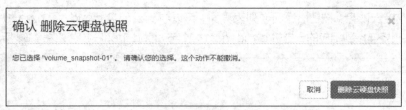

图 7.35　删除云硬盘快照

在 Dashboard 界面中，先单击"项目"主节点，再单击"管理员"→"镜像"子节点，打开镜像列表界面，如图 7.36 所示，可以查看镜像列表信息。

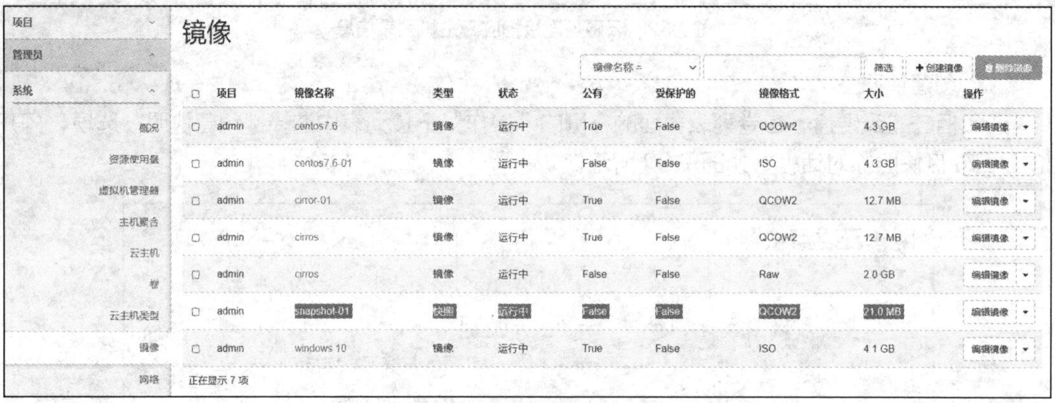

图 7.36　查看镜像列表信息

（8）修改云硬盘类型

选择相应的云硬盘，在卷列表界面的右侧的"操作"下拉列表中选择"修改云硬盘类型"选项，打开"修改云硬盘类型"对话框，在卷创建后更改卷类型。其等同于命令 cinder retype 的操作。选定的"卷类型"必须与当前的卷类型不同。"迁移策略"只在云硬盘更改类型无法完成时使用。如果"迁移策略"是"按需"，后端会执行云硬盘迁移。注意，迁移可能会占用大量的时间来完成，有时可达数小时，如图 7.37 所示。

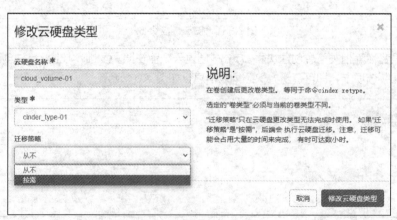

图 7.37　修改云硬盘类型

（9）上传云硬盘到镜像

选择相应的云硬盘，在卷列表界面的右侧的"操作"下拉列表中选择"上传镜像"选项，打开

"上传云硬盘到镜像"对话框。上传卷到镜像服务,作为镜像。其等同于命令 cinder upload-to-image 的操作,即为镜像选择"磁盘格式"。卷镜像被 QEMU 磁盘镜像工具创建。当卷状态是"使用中"时,可以用"强制"来上传卷到一个镜像中,如图 7.38 所示。

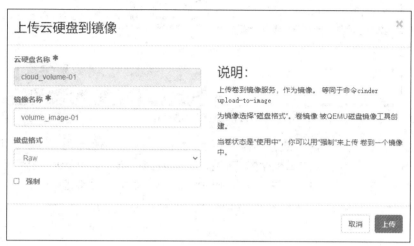

图 7.38 上传云硬盘到镜像

(10)管理员系统中的云硬盘管理

在 Dashboard 界面中,先单击"管理员"主节点,再单击"系统"→"卷"子节点,打开项目中的卷列表界面,如图 7.39 所示,可以查看项目中的卷列表信息。

图 7.39 查看项目中的卷列表信息

在项目中的"卷"选项卡右侧的"操作"下拉列表中选择"删除云硬盘"选项,可以删除云硬盘,如图 7.40 所示。

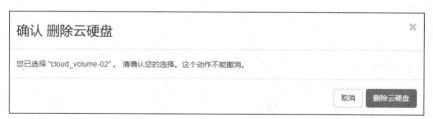

图 7.40 删除云硬盘

在项目中的"卷"选项卡右侧的"操作"下拉列表中选择"更新云硬盘状态"选项,可以更新云硬盘状态,如图 7.41 所示。云硬盘状态通常是自动管理的。在某些特定的环境下,管理员需要手动更新云硬盘状态。其等同于命令 cinder snapshot-reset-state 的操作。

图 7.41 更新云硬盘状态

在项目中的"卷"选项卡右侧的"操作"下拉列表中选择"非管理云硬盘"选项,可以不管理云硬盘,如图 7.42 所示。当某个云硬盘被"非管理"时,它将不在 OpenStack 中可见。注意,这并不是在 Cinder 主机中将云硬盘删除。其等同于 cinder unmanage 命令的操作。

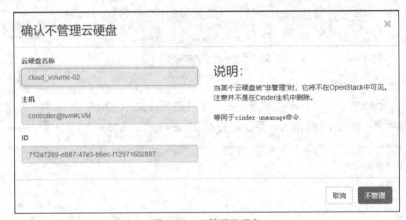

图 7.42 不管理云硬盘

在项目中的"卷"选项卡右侧的"操作"下拉列表中,选择"迁移卷"选项,可以迁移卷,如图 7.43 所示。其中,可以热迁移一个云主机到指定主机;而强制主机复制表示启用或禁用通用的基于主机的强制迁移,跳过驱动器优化。

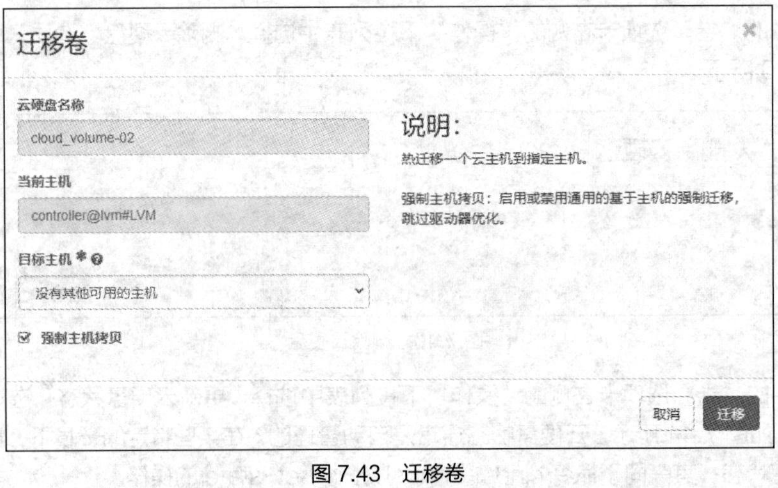

图 7.43 迁移卷

在项目中的卷列表界面中,选择"云硬盘类型"选项卡,查看云硬盘类型,如图 7.44 所示。

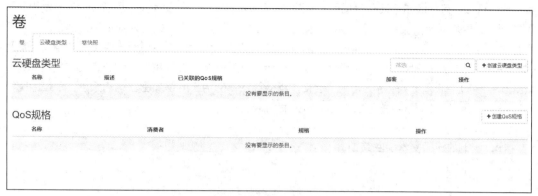

图 7.44　查看云硬盘类型

在"云硬盘类型"选项卡右侧,单击"+创建云硬盘类型"按钮,打开"创建云硬盘类型"对话框,如图 7.45 所示。云硬盘类型是在创建云硬盘的时候可以指定的标签。它通常用来映射到这些云硬盘所使用的存储后端驱动器的性能指标集合,如"性能""SSD""备份"等。这等同于 cinder type-create 命令的操作。一旦云硬盘类型被创建,单击"查看扩展规格"按钮就可以设置云硬盘类型的扩展规格键值对。

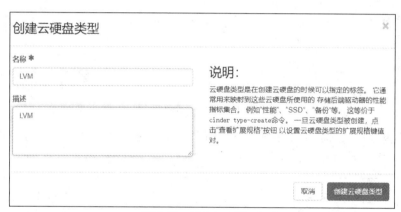

图 7.45　创建云硬盘类型

单击"创建云硬盘类型"按钮,返回云硬盘类型列表界面,如图 7.46 所示。

图 7.46　云硬盘类型列表界面

在"云硬盘类型"选项卡中,选择相应的云硬盘类型,在右侧的"操作"下拉列表中选择"创建加密"选项,可创建加密云硬盘类型,如图7.47所示。为一个云硬盘类型创建加密后,使用此类型的所有云硬盘都会被加密。若云硬盘类型当前正在被云硬盘使用,则无法添加加密信息。提供者是提供加密支持的 class(如 LuksEncryptor)。控制位置用于指定由哪个服务来对云硬盘进行加密(如"前端"对应的是 Nova)。其默认值是"前端"。Cipher 是要用的加密算法或方式(如 aes-xts-plain64)。如果该字段空白,则将会使用默认算法。密钥长度是加密密钥的大小,单位为比特(如 128、256)。如果该字段为空白,则将会使用默认值。

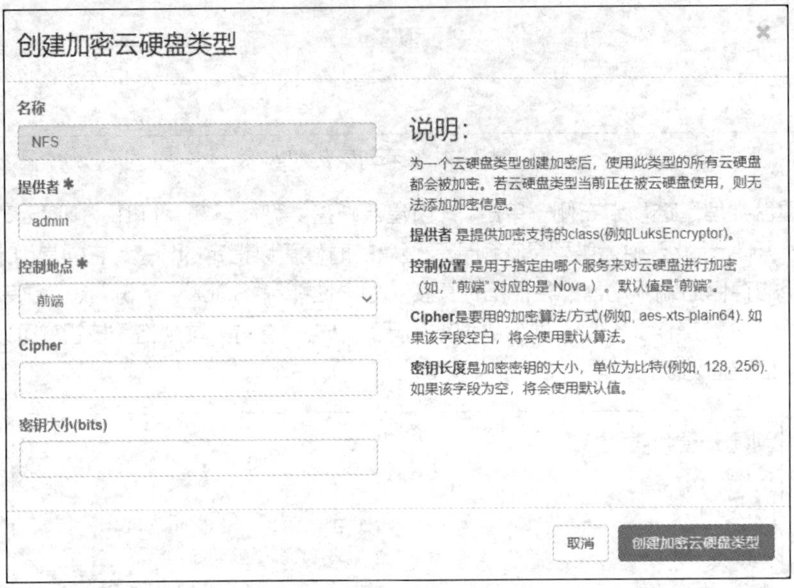

图 7.47 创建加密云硬盘类型

在"云硬盘类型"选项卡中,选择相应的云硬盘类型,在右侧的"操作"下拉列表中选择"查看扩展规格"选项,可查看云硬盘类型扩展规格,如图 7.48 所示。

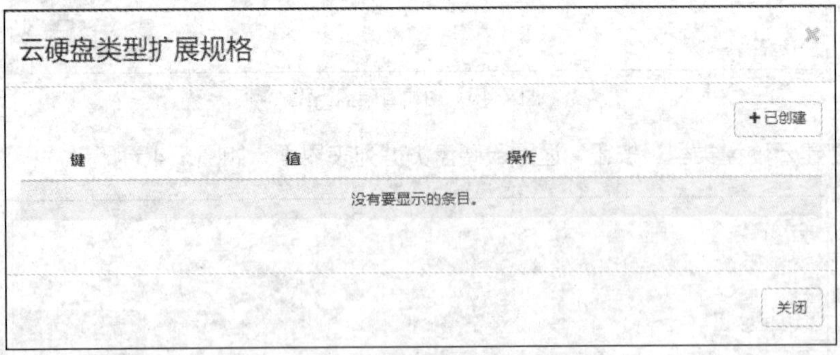

图 7.48 查看云硬盘类型扩展规格

在"云硬盘类型扩展规格"对话框中,单击右侧的"+已创建"按钮,打开"创建云硬盘类型扩展规格"对话框,如图 7.49 所示。

在"创建云硬盘类型扩展规格"对话框中,为云硬盘类型创建新的"扩展规格"键值对,输入相关数据,单击"+已创建"按钮,完成创建云硬盘类型扩展规格操作,打开"云硬盘类型:NFS"对话框,如图 7.50 所示。

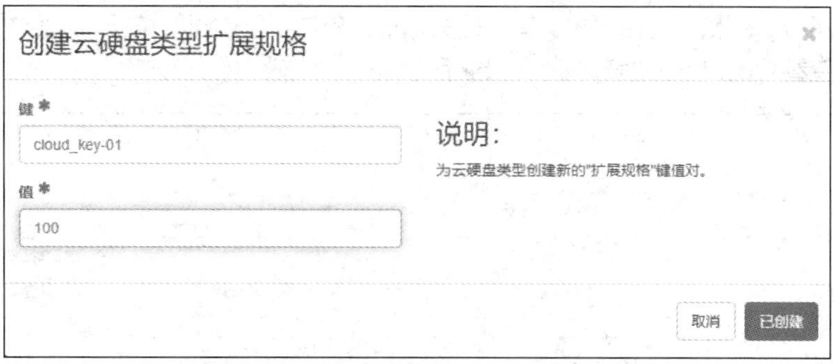

图 7.49 "创建云硬盘类型扩展规格"对话框

图 7.50 "云硬盘类型：NFS"对话框

在"云硬盘类型：NFS"对话框中，单击"关闭"按钮，返回项目中的卷列表界面，在 QoS 界面右侧单击"创建 QoS 规格"按钮，打开"创建 QoS 规格"对话框，如图 7.51 所示。QoS 规格可以和卷类型进行关联。它用于对卷所属者所请求的服务质量指标进行限制。这等同于 cinder qos-create 命令的操作。当 QoS 规格创建成功之后，单击"管理规格"按钮可对 QoS 规格中的规格键值对进行修改。每一个 QoS 规格都有一个"消费者"的属性，代表着 QoS 策略的实施对象。该属性可以设置为"前端"（对应 Nova 计算服务）、"后端"（对应 Cinder 后端）以及"前后两端"。

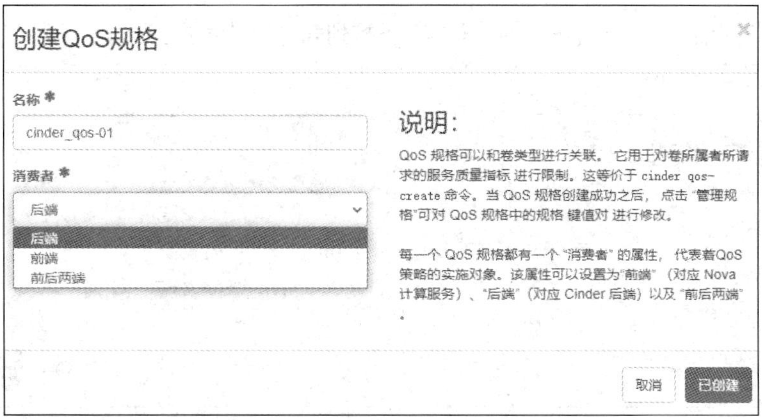

图 7.51 "创建 QoS 规格"对话框

在"创建 QoS 规格"对话框中,输入相关信息,单击"已创建"按钮,返回"云硬盘类型"选项卡,即可查看 QoS 规格列表,如图 7.52 所示。

图 7.52 查看 QoS 规格列表

在"QoS 规格"选项组中,选择相应的 QoS 规格名称,选择右侧的"操作"下拉列表中的"管理规格"选项,打开"规格"对话框,如图 7.53 所示。

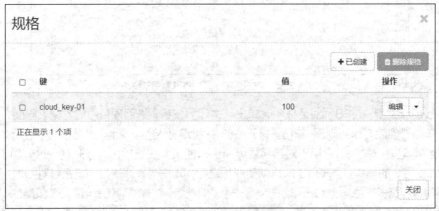

图 7.53 "规格"对话框

在"规格"对话框中,选择右侧的"操作"下拉列表中的"编辑"选项,打开"编辑规格的值"对话框,如图 7.54 所示。

图 7.54 "编辑规格的值"对话框

输入相应值,单击"保存"按钮,返回规格界面,在"QoS 规格"选项组中,在其右侧的"操作"下拉列表中选择"编辑消费者"选项,打开"编辑 QoS 规格的消费者"对话框,如图 7.55 所示。每个 QoS 规格实体都有一个"消费者"的值,指明管理员希望 QoS 策略实施的位置。该值可以是"前端"(Nova 计算服务)、"后端"(Cinder 后端)或者"前后两端"。

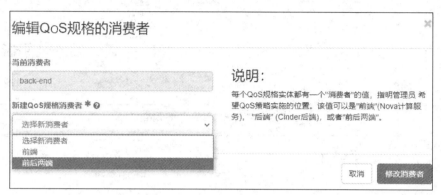

图 7.55 "编辑 QoS 规格的消费者"对话框

在"云硬盘类型"选项卡中,选择相应的云硬盘,在"操作"下拉列表中选择"管理 QoS 规格关联"选项,打开"将 QoS 规格关联到云硬盘类型"对话框,如图 7.56 所示。可以添加、修改、删除该云硬盘类型所关联的 QoS 规格。"无"表示当前没有关联 QoS 规格。同样的,将 QoS 规格设置为"无"会解除当前的关联。这等同于 cinder qos-associate 和 cinder qos-disassociate 命令的操作。

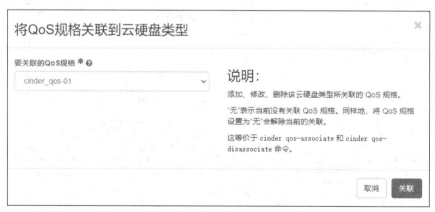

图 7.56 "将 QoS 规格关联到云硬盘类型"对话框

在"将 QoS 规格关联到云硬盘类型"对话框中,单击"关联"按钮,完成将 QoS 规格关联到云硬盘类型操作。返回"云硬盘类型"选项卡,可以查看已关联的 QoS 规格,如图 7.57 所示。

2. 对象存储管理

(1)查看容器列表信息

在 Dashboard 界面中,先单击"项目"主节点,再单击"对象存储"→"容器"子节点,打开容器列表界面,如图 7.58 所示,可以查看容器列表信息。

(2)创建容器及文件夹

在容器列表信息界面中,单击"+容器"按钮,打开"创建容器"对话框,如图 7.59 所示。注意:公有容器允许任何人通过公共 URL 来获取容器中的对象。

图 7.57　查看已关联的 QoS 规格

图 7.58　查看容器列表信息

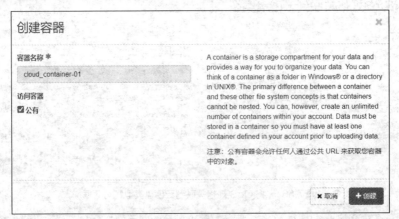

图 7.59　"创建容器"对话框

在"创建容器"对话框中，单击"+创建"按钮，完成容器 cloud_container-01 的创建，如图 7.60 所示。

在容器 cloud_container-01 中创建文件夹，单击"+文件夹"按钮，打开"Create Folder In: cloud_container-01"对话框，如图 7.61 所示。

> **注意**
> 允许用文件夹名称中的分隔符（'/'）来创建深层文件夹。

图 7.60 完成容器 cloud_container-01 的创建

图 7.61 在容器 cloud_container-01 中创建文件夹

在"Create Folder In: cloud_container-01"对话框中,输入文件夹名,单击"创建文件夹"按钮,完成文件夹的创建,如图 7.62 所示。

图 7.62 在容器 cloud_container-01 中显示创建的文件夹 file-01

(3)上传文件

选择容器 cloud_container-01 的文件夹 file-01,上传文件到文件夹 file-01 中,如图 7.63 所示。

图 7.63 上传文件到文件夹 file-01 中

在容器 cloud_container-01 的文件夹 file-01 中,单击"上传文件"按钮,打开"Upload File To: cloud_container-01:file-01"对话框,单击"选择文件"按钮,选择需要上传的文件,如图 7.64 所示。

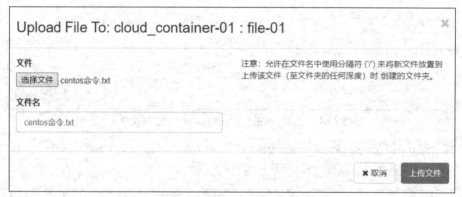

图 7.64 选择需要上传的文件

单击"上传文件"按钮,完成上传文件到容器的文件夹中的操作,如图 7.65 所示。

图 7.65 完成上传文件到容器的文件夹中的操作

（4）下载文件

在容器 cloud_container-01 的文件夹 file-01 中，选择要下载的文件，选择"下载"选项，如图 7.66 所示，可以下载文件到本地磁盘。

图 7.66　下载文件到本地磁盘

单击"全部显示"按钮，可以显示全部文件的下载情况，如图 7.67 所示。

图 7.67　显示全部文件的下载情况

选择"在文件夹中显示"选项，可以查看文件的详细信息，如图 7.68 所示。

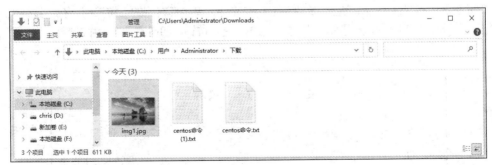

图 7.68　查看文件的详细信息

（5）容器中文件及文件夹的管理

查看文件的详细信息。选择容器中文件夹的相应文件，单击其右侧的下拉按钮，在下拉列表中选择"查看详细信息"选项，查看文件的详细信息，如图7.69所示。

图7.69　查看文件的详细信息

例如，选择文件centos命令.txt进行查看操作，选择"查看详细信息"选项，打开"对象详细信息"对话框，如图7.70所示。

图7.70　"对象详细信息"对话框

删除容器中的文件。选择容器中文件夹的相应文件，单击其右侧的下拉按钮，在下拉列表中选择"删除"选项，打开"确认删除"对话框，单击"是"按钮，即可删除容器中的文件，如图7.71所示。

图7.71　删除容器中的文件

删除容器中的文件夹。选择相应的容器中的文件夹，单击其右侧的下拉按钮，在下拉列表中选择"删除"选项，如图 7.72 所示。

图 7.72　删除容器中的文件夹

例如，选择容器中的文件夹 image-01 进行删除操作，选择"删除"选项，打开"确认删除"对话框，如图 7.73 所示，单击"是"按钮，即可删除文件夹。

图 7.73　"确认删除"对话框

删除容器中的文件夹时，必须知道文件夹中有无文件，如果有，则文件夹无法删除，如图 7.74 所示。

图 7.74　无法删除容器中的非空文件夹

7.3.2　基于命令行界面管理存储服务

系统管理员一般会基于命令行来管理存储服务，可以使用 openstack 命令或 cinder 命令管理块存储，使用 openstack 命令或 swift 命令管理对象存储。

1. 块存储管理

可使用 openstack 命令或 cinder 命令管理块存储，使用帮助命令可以查看块存储的相关命令，执行命令如下。

```
# openstack help | grep volume
```

或者

```
# cinder help | grep volume
```

其命令的结果如下。

```
[root@controller ~]# openstack help | grep volume
                    [--os-volume-api-version <volume-api-version>]
  --os-volume-api-version <volume-api-version>
  server add volume    Add volume to server
  server remove volume   Remove volume from server
  snapshot delete    Delete volume snapshot(s)
  volume create    Create new volume
  volume delete    Delete volume(s)
  volume list      List volumes
  volume qos associate   Associate a QoS specification to a volume type
  volume qos create    Create new QoS specification
  volume qos delete    Delete QoS specification
  volume qos disassociate   Disassociate a QoS specification from a volume type
  volume qos list    List QoS specifications
  volume qos set     Set QoS specification properties
  volume qos show    Display QoS specification details
  volume qos unset   Unset QoS specification properties
  volume set       Set volume properties
  volume show      Display volume details
  volume type create   Create new volume type
  volume type delete   Delete volume type
  volume type list   List volume types
  volume type set    Set volume type properties
  volume type show   Display volume type details
  volume type unset   Unset volume type properties
  volume unset     Unset volume properties
[root@controller ~]#
[root@controller ~]# cinder help | grep volume
                    [--volume-service-name <volume-service-name>]
                    [--os-volume-api-version <volume-api-ver>]
    backup-create       Creates a volume backup.
    create              Creates a volume.
    delete              Removes one or more volumes.
                        Creates encryption type for a volume type. Admin only.
                        Deletes encryption type for a volume type. Admin only.
                        Shows encryption type details for volume types. Admin
                        Shows encryption type details for a volume type. Admin
                        Update encryption type information for a volume type
    extend              Attempts to extend size of an existing volume.
```

```
    extra-specs-list    Lists current volume types and extra specs.
    force-delete        Attempts force-delete of volume, regardless of state.
    get-capabilities    Show backend volume stats and properties. Admin only.
    image-metadata      Sets or deletes volume image metadata.
                        Shows volume image metadata.
    list                Lists all volumes.
    manage              Manage an existing volume.
    metadata            Sets or deletes volume metadata.
    metadata-show       Shows volume metadata.
                        Updates volume metadata.
    migrate             Migrates volume to a new host.
    qos-associate       Associates qos specs with specified volume type.
    qos-disassociate    Disassociates qos specs from specified volume type.
                        Updates volume read-only access-mode flag.
    rename              Renames a volume.
                        Promote a secondary volume to primary for a
                        Sync the secondary volume with primary for a
    reset-state         Explicitly updates the volume state in the Cinder
    retype              Changes the volume type for a volume.
    set-bootable        Update bootable status of a volume.
    show                Shows volume details.
    transfer-accept     Accepts a volume transfer.
    transfer-create     Creates a volume transfer.
    type-access-add     Adds volume type access for the given project.
    type-access-list    Print access information about the given volume type.
    type-access-remove  Removes volume type access for the given project.
    type-create         Creates a volume type.
    type-default        List the default volume type.
    type-delete         Deletes a volume type.
    type-key            Sets or unsets extra_spec for a volume type.
    type-list           Lists available 'volume types'. (Admin only will see
    type-show           Show volume type details.
    type-update         Updates volume type name, description, and/or
    unmanage            Stop managing a volume.
    upload-to-image     Uploads volume to Image Service as an image.
                        Service type. For most actions, default is volume.
  --volume-service-name <volume-service-name>
  --os-volume-api-version <volume-api-ver>
[root@controller ~]#
```

（1）查看块存储卷列表信息

查看块存储卷列表信息可以使用 openstack 命令，也可以使用 cinder 命令，执行命令如下。

```
#openstack  volume  list
```

或者

```
#cinder  list
```

其命令的结果如图 7.75 所示。

```
[root@controller ~]# openstack volume list
+--------------------------------------+----------------+--------+------+------------------------------------------+
| ID                                   | Display Name   | Status | Size | Attached to                              |
+--------------------------------------+----------------+--------+------+------------------------------------------+
| 712a7289-e887-47e3-b6ec-f12971602887 | cloud_volume-02| in-use |   3  | Attached to cloud_host-10 on /dev/vdb    |
| 304db00f-3a89-491d-8708-71545606d708 | cloud_volume-01| in-use |   1  | Attached to cloud_host-01 on /dev/vdb    |
+--------------------------------------+----------------+--------+------+------------------------------------------+
[root@controller ~]# cinder list
+--------------------------------------+--------+----------------+------+-------------+----------+--------------------------------------+
|                  ID                  | Status |      Name      | Size | Volume Type | Bootable |             Attached to              |
+--------------------------------------+--------+----------------+------+-------------+----------+--------------------------------------+
| 304db00f-3a89-491d-8708-71545606d708 | in-use | cloud_volume-01|  1   |      -      |  false   | 5282ca6c-f232-45e8-8fee-5cd94ab6fcda |
| 712a7289-e887-47e3-b6ec-f12971602887 | in-use | cloud_volume-02|  3   |      -      |  true    | 6e9402bf-9f1c-402b-bd6c-8ec711af8c5d |
+--------------------------------------+--------+----------------+------+-------------+----------+--------------------------------------+
[root@controller ~]#
```

图 7.75 查看块存储卷列表信息

（2）查看块存储卷类型列表信息

查看块存储卷类型列表信息可以使用 openstack 命令，也可以使用 cinder 命令，执行命令如下。

```
# openstack  volume  type  list
```

或者

```
# cinder  type-list
```

其命令的结果如图 7.76 所示。

```
[root@controller ~]# openstack volume type list
+--------------------------------------+------+
| ID                                   | Name |
+--------------------------------------+------+
| e24c4212-e23b-46c8-990b-6e6818d8fe9f | NFS  |
| 8f4953d5-37eb-4a37-b48e-b691134f8235 | LVM  |
+--------------------------------------+------+
[root@controller ~]# cinder type-list
+--------------------------------------+------+-------------+-----------+
|                  ID                  | Name | Description | Is_Public |
+--------------------------------------+------+-------------+-----------+
| 8f4953d5-37eb-4a37-b48e-b691134f8235 | LVM  |     LVM     |    True   |
| e24c4212-e23b-46c8-990b-6e6818d8fe9f | NFS  |     NFS     |    True   |
+--------------------------------------+------+-------------+-----------+
[root@controller ~]#
```

图 7.76 查看块存储卷类型列表信息

（3）创建卷类型

创建卷类型可以使用 openstack 命令，也可以使用 cinder 命令，使用帮助命令可以查看创建卷类型的相关命令的具体参数，执行命令如下。

```
# openstack  help  volume  type  create
```

或者

```
# cinder help type-create
```

其命令的结果如下。

```
[root@controller ~]# openstack help volume type create
usage: openstack volume type create [-h]
                                    [-f {html,json,json,shell,table,value,yaml,yaml}]
                                    [-c COLUMN] [--max-width <integer>]
                                    [--noindent] [--prefix PREFIX]
                                    [--description <description>]
                                    [--public | --private]
                                    [--property <key=value>]
                                    <name>

Create new volume type
positional arguments:
  <name>                New volume type name
optional arguments:
  -h, --help            show this help message and exit
```

```
  --description <description>
                        New volume type description
  --public              Volume type is accessible to the public
  --private             Volume type is not accessible to the public
  --property <key=value>
                        Property to add for this volume type(repeat option to
                        set multiple properties)
[root@controller ~]#
[root@controller ~]# cinder help type-create
usage: cinder type-create [--description <description>]
                          [--is-public <is-public>]
                          <name>

Creates a volume type.
Positional arguments:
  <name>                Name of new volume type.
Optional arguments:
  --description <description>
                        Description of new volume type.
  --is-public <is-public>
                        Make type accessible to the public (default true).
[root@controller ~]#
```

例如，使用 openstack 命令创建一个指定的公共卷类型 lvm，执行如下命令。

```
# openstack volume type create --public lvm
```

其命令的结果如图 7.77 所示。

```
[root@controller ~]# openstack volume type create --public lvm
+-----------------------------+--------------------------------------+
| Field                       | Value                                |
+-----------------------------+--------------------------------------+
| description                 | None                                 |
| id                          | dd255b7e-5f50-45eb-b152-7b125dcef833 |
| is_public                   | True                                 |
| name                        | lvm                                  |
| os-volume-type-access:is_public | True                             |
+-----------------------------+--------------------------------------+
[root@controller ~]# openstack volume type list
+--------------------------------------+------+
| ID                                   | Name |
+--------------------------------------+------+
| dd255b7e-5f50-45eb-b152-7b125dcef833 | lvm  |
+--------------------------------------+------+
[root@controller ~]#
```

图 7.77 使用 openstack 命令创建一个指定的公共卷类型 lvm

又如，使用 cinder 命令创建一个指定的公共卷类型 nfs，执行如下命令。

```
# cinder type-create --is-public True nfs
```

其命令的结果如图 7.78 所示。

```
[root@controller ~]# cinder type-create --is-public True nfs
+--------------------------------------+------+-------------+-----------+
| ID                                   | Name | Description | Is_Public |
+--------------------------------------+------+-------------+-----------+
| fae365f6-352a-4652-91c0-e7f085aeaa79 | nfs  | -           | True      |
+--------------------------------------+------+-------------+-----------+
[root@controller ~]# cinder type-list
+--------------------------------------+------+-------------+-----------+
| ID                                   | Name | Description | Is_Public |
+--------------------------------------+------+-------------+-----------+
| dd255b7e-5f50-45eb-b152-7b125dcef833 | lvm  | -           | True      |
| fae365f6-352a-4652-91c0-e7f085aeaa79 | nfs  | -           | True      |
+--------------------------------------+------+-------------+-----------+
[root@controller ~]#
```

图 7.78 使用 cinder 命令创建一个指定的公共卷类型 nfs

（4）删除卷类型

例如，使用 openstack 命令删除卷类型 lvm，执行如下命令。

openstack volume type delete vlm

其命令的结果如图 7.79 所示。

```
[root@controller ~]# openstack volume type delete lvm
[root@controller ~]# openstack volume type list
+--------------------------------------+------+
| ID                                   | Name |
+--------------------------------------+------+
| fae365f6-352a-4652-91c0-e7f085aeaa79 | nfs  |
+--------------------------------------+------+
[root@controller ~]#
```

图 7.79　使用 openstack 命令删除卷类型 lvm

又如，使用 cinder 命令删除卷类型 nfs，执行如下命令。

cinder type-delete nfs

其命令的结果如图 7.80 所示。

```
[root@controller ~]# cinder type-delete fae365f6-352a-4652-91c0-e7f085aeaa79
[root@controller ~]# cinder type-list
+----+------+-------------+-----------+
| ID | Name | Description | Is_Public |
+----+------+-------------+-----------+
+----+------+-------------+-----------+
[root@controller ~]#
```

图 7.80　使用 cinder 命令删除卷类型 nfs

（5）创建卷

创建卷可以使用 openstack 命令，也可以使用 cinder 命令，使用帮助命令可以查看创建卷的相关命令的具体参数，执行如下命令。

openstack help volume create

或者

cinder help create

其命令的结果如下。

```
[root@controller ~]# openstack help volume create
usage: openstack volume create [-h]
                               [-f {html,json,json,shell,table,value,yaml,yaml}]
                               [-c COLUMN] [--max-width <integer>]
                               [--noindent] [--prefix PREFIX] --size <size>
                               [--snapshot <snapshot>]
                               [--description <description>]
                               [--type <volume-type>] [--user <user>]
                               [--project <project>]
                               [--availability-zone <availability-zone>]
                               [--image <image>] [--source <volume>]
                               [--property <key=value>]
                               <name>

Create new volume
positional arguments:
    <name>                 New volume name
[root@controller ~]#
[root@controller ~]# cinder help create
```

```
usage: cinder create [--consisgroup-id <consistencygroup-id>]
                     [--snapshot-id <snapshot-id>]
                     [--source-volid <source-volid>]
                     [--source-replica <source-replica>]
                     [--image-id <image-id>] [--image <image>] [--name <name>]
                     [--description <description>]
                     [--volume-type <volume-type>]
                     [--availability-zone <availability-zone>]
                     [--metadata [<key=value> [<key=value> ...]]]
                     [--hint <key=value>] [--allow-multiattach]
                     [<size>]
Creates a volume.
Positional arguments:
    <size>              Size of volume, in GiBs. (Required unless snapshot-id
                        /source-volid is specified).
[root@controller ~]#
```

例如，使用 openstack 命令创建一个容量大小为 2GB、卷类型为 lvm、卷名为 cloud_volume-10 的云硬盘，执行如下命令。

```
# openstack volume create --size 2 --type lvm cloud_volume-10
```

其命令的结果如图 7.81 所示。

```
[root@controller ~]# openstack volume create --size 2 --type lvm cloud_volume-10
+---------------------+--------------------------------------+
| Field               | Value                                |
+---------------------+--------------------------------------+
| attachments         | []                                   |
| availability_zone   | nova                                 |
| bootable            | false                                |
| consistencygroup_id | None                                 |
| created_at          | 2021-01-25T20:12:26.810471           |
| description         | None                                 |
| encrypted           | False                                |
| id                  | f5fd357c-2add-476b-b748-01d01f980aa6 |
| migration_status    | None                                 |
| multiattach         | False                                |
| name                | cloud_volume-10                      |
| properties          |                                      |
| replication_status  | disabled                             |
| size                | 2                                    |
| snapshot_id         | None                                 |
| source_volid        | None                                 |
| status              | creating                             |
| type                | lvm                                  |
| updated_at          | None                                 |
| user_id             | 0befa70f767848e39df8224107b71858     |
+---------------------+--------------------------------------+
[root@controller ~]#
```

图 7.81　使用 openstack 命令创建云硬盘 cloud_volume-10

又如，使用 cinder 命令创建一个容量大小为 2GB、卷类型为 nfs、卷名为 cloud_volume-20 的云硬盘，执行如下命令。

```
# cinder create --name cloud_volume-20 --volume-type nfs 2
```

其命令的结果如图 7.82 所示。

（6）删除卷

删除卷可以使用 openstack 命令，也可以使用 cinder 命令，使用帮助命令可以查看删除卷的相关命令的具体参数，执行命令的格式如下。

```
# openstack volume delete 卷名称
```

或者

```
# cinder delete 卷名称
```

```
[root@controller ~]# cinder create --name cloud_volume-20 --volume-type nfs 2
+---------------------------------+--------------------------------------+
|            Property             |                Value                 |
+---------------------------------+--------------------------------------+
|           attachments           |                  []                  |
|        availability_zone        |                 nova                 |
|             bootable            |                false                 |
|       consistencygroup_id       |                 None                 |
|            created_at           |       2021-01-25T20:16:15.000000     |
|           description           |                 None                 |
|            encrypted            |                False                 |
|               id                | aa862952-4dbc-439b-88a5-9be5682fea2d |
|            metadata             |                  {}                  |
|         migration_status        |                 None                 |
|           multiattach           |                False                 |
|              name               |           cloud_volume-20            |
|       os-vol-host-attr:host     |                 None                 |
|  os-vol-mig-status-attr:migstat |                 None                 |
| os-vol-mig-status-attr:name_id  |                 None                 |
|    os-vol-tenant-attr:tenant_id |   f9ff39ba9daa4e5a8fee1fc50e2d2b34   |
|       replication_status        |               disabled               |
|              size               |                  2                   |
|           snapshot_id           |                 None                 |
|          source_volid           |                 None                 |
|             status              |               creating               |
|           updated_at            |                 None                 |
|             user_id             |   0befa70f767848e39df8224107b71858   |
|           volume_type           |                 nfs                  |
+---------------------------------+--------------------------------------+
[root@controller ~]#
```

图 7.82　使用 cinder 命令创建云硬盘 cloud_volume-20

其命令的结果如图 7.83 所示。

```
[root@controller ~]# openstack volume delete cloud_volume-30
[root@controller ~]# openstack volume list
+--------------------------------------+-----------------+-----------+------+-----------------------------------+
| ID                                   | Display Name    | Status    | Size | Attached to                       |
+--------------------------------------+-----------------+-----------+------+-----------------------------------+
| 712a7289-e887-47e3-b6ec-f12971602887 | cloud_volume-02 | in-use    | 3    | Attached to cloud_host-10 on /dev/vdb |
| 304db00f-3a89-491d-8708-71545606d708 | cloud_volume-01 | available | 1    |                                   |
+--------------------------------------+-----------------+-----------+------+-----------------------------------+
[root@controller ~]#

[root@controller ~]# cinder delete  cloud_volume-30
Request to delete volume cloud_volume-30 has been accepted.
[root@controller ~]# cinder list
+--------------------------------------+-----------+-----------------+------+-------------+----------+--------------------------------------+
|                  ID                  |   Status  |       Name      | Size | Volume Type | Bootable |             Attached to              |
+--------------------------------------+-----------+-----------------+------+-------------+----------+--------------------------------------+
| 304db00f-3a89-491d-8708-71545606d708 | available | cloud_volume-01 |  1   |      -      |  false   |                                      |
| 712a7289-e887-47e3-b6ec-f12971602887 |  in-use   | cloud_volume-02 |  3   |      -      |   true   | 6e9402bf-9f1c-402b-bd6c-8ec711af8c5d |
+--------------------------------------+-----------+-----------------+------+-------------+----------+--------------------------------------+
[root@controller ~]#
```

图 7.83　使用 openstack 命令与 cinder 命令删除卷

（7）云硬盘连接云主机

使用命令行工具使云硬盘连接到云主机时，需要先查看当前可用的云硬盘与主机的使用情况，使用 cinder 命令与 nova 命令查看当前云硬盘与云主机列表信息，执行如下命令。

[root@controller ~]#cinder　　list
[root@controller ~]#nova　　list

其命令的结果如图 7.84 所示。

```
[root@controller ~]# cinder list
+--------------------------------------+-----------+-----------------+------+-------------+----------+--------------------------------------+
|                  ID                  |   Status  |       Name      | Size | Volume Type | Bootable |             Attached to              |
+--------------------------------------+-----------+-----------------+------+-------------+----------+--------------------------------------+
| 304db00f-3a89-491d-8708-71545606d708 | available | cloud_volume-01 |  1   |      -      |  false   |                                      |
| 712a7289-e887-47e3-b6ec-f12971602887 |  in-use   | cloud_volume-02 |  3   |      -      |   true   | 6e9402bf-9f1c-402b-bd6c-8ec711af8c5d |
+--------------------------------------+-----------+-----------------+------+-------------+----------+--------------------------------------+
[root@controller ~]#
[root@controller ~]# nova list
+--------------------------------------+---------------+--------+------------+-------------+------------------------------------------+
| ID                                   | Name          | Status | Task State | Power State | Networks                                 |
+--------------------------------------+---------------+--------+------------+-------------+------------------------------------------+
| 5282ca6c-f232-45e8-8fee-5cd94ab6fcda | cloud_host-01 | ACTIVE | -          | Running     | in-net=192.168.100.101, 202.199.184.102  |
| 6e9402bf-9f1c-402b-bd6c-8ec711af8c5d | cloud_host-10 | ACTIVE | -          | Running     | in-net=192.168.100.103                   |
| 27a462e8-60ef-4d1b-87dd-5a63a22fb300 | cloud_host-20 | ACTIVE | -          | Running     | out-net=202.199.184.105                  |
+--------------------------------------+---------------+--------+------------+-------------+------------------------------------------+
[root@controller ~]#
```

图 7.84　查看当前云硬盘与云主机列表信息

从图 7.84 中可以看出，云硬盘 cloud_volume-01 没有连接到任何云主机，云主机 cloud_host-01 处于运行状态，这里将云硬盘 cloud_volume-01 连接到云主机 cloud_host-01。通过 URL 的方式，访问云主机 cloud_host-01，使用 nova 命令查看云主机的 URL 信息，执行如下命令。

[root@controller ~]# nova get-vnc-console cloud_host-01 novnc

其命令的结果如图 7.85 所示。

```
[root@controller ~]# nova get-vnc-console cloud_host-01 novnc
+-------+------------------------------------------------------------------------------+
| Type  | Url                                                                          |
+-------+------------------------------------------------------------------------------+
| novnc | http://192.168.100.10:6080/vnc_auto.html?token=1fd603d7-29d0-4518-bb00-a14909907a7d |
+-------+------------------------------------------------------------------------------+
[root@controller ~]#
```

图 7.85 查看云主机的 URL 信息

在 Web 浏览器的地址栏中输入 URL 信息，可以访问云主机 cloud_host-01，输入用户名为 cirros，密码为 cubswin:)，进行登录访问，如图 7.86 所示。

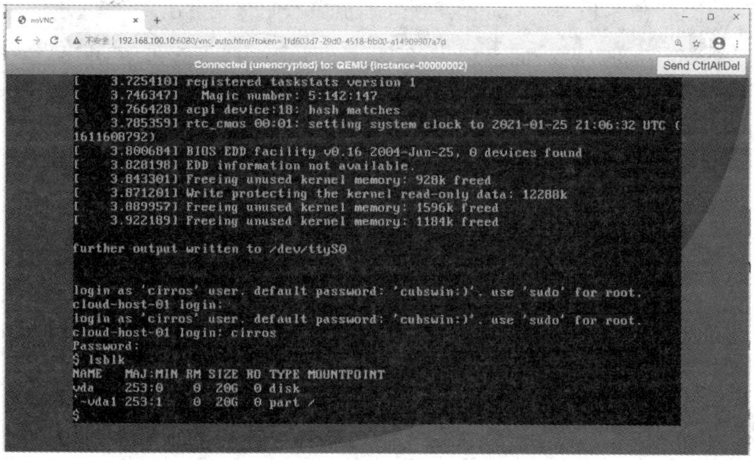

图 7.86 Web 浏览器方式访问云主机 cloud_host-01

在云主机上输入 lsblk 命令，可以查看当前磁盘分区情况，发现其只有一块 vda1 磁盘，磁盘盘符以字母"v"开始，说明这块磁盘是网络云磁盘，容量为 20GB。

使用帮助命令，查看云主机连接云硬盘所使用的命令，执行如下命令。

```
[root@controller ~]# nova help | grep volume
            [--volume-service-name <volume-service-name>]
    volume-attach               Attach a volume to a server.
    volume-attachments          List all the volumes attached to a server.
    volume-create               DEPRECATED: Add a new volume.
    volume-delete               DEPRECATED: Remove volume(s).
    volume-detach               Detach a volume from a server.
    volume-list                 DEPRECATED: List all the volumes.
    volume-show                 DEPRECATED: Show details about a volume.
    volume-snapshot-create      DEPRECATED: Add a new snapshot.
    volume-snapshot-delete      DEPRECATED: Remove a snapshot.
    volume-snapshot-list        DEPRECATED: List all the snapshots.
    volume-snapshot-show        DEPRECATED: Show details about a snapshot.
    volume-type-create          DEPRECATED: Create a new volume type.
    volume-type-delete          DEPRECATED: Delete a specific volume type.
    volume-type-list            DEPRECATED: Print a list of available 'volume
    volume-update               Update volume attachment.
```

```
    --volume-service-name <volume-service-name>
[root@controller ~]#
```

通过帮助命令可以查到，使用 nova volume-attach 命令可以进行云硬盘与云主机的连接，继续使用帮助命令对 volume-attach 命令参数进行解析，执行如下命令。

```
[root@controller ~]# nova help volume-attach
usage: nova volume-attach <server> <volume> [<device>]
Attach a volume to a server.
Positional arguments:
  <server>    Name or ID of server.
  <volume>    ID of the volume to attach.
  <device>    Name of the device e.g. /dev/vdb. Use "auto" for autoassign (if
              supported). Libvirt driver will use default device name.
[root@controller ~]#
```

可以看出使用 nova volume-attach 命令时，连接云主机可以用云主机名称或 ID，但连接云硬盘必须用云硬盘的 ID，执行如下命令。

```
# nova volume-attach  cloud_host-01  304db00f-3a89-491d-8708-71545606d708
```

其命令的结果如图 7.87 所示。

```
[root@controller ~]# nova volume-attach  cloud_host-01  304db00f-3a89-491d-8708-71545606d708
+----------+--------------------------------------+
| Property | Value                                |
+----------+--------------------------------------+
| device   | /dev/vdb                             |
| id       | 304db00f-3a89-491d-8708-71545606d708 |
| serverId | 5282ca6c-f232-45e8-8fee-5cd94ab6fcda |
| volumeId | 304db00f-3a89-491d-8708-71545606d708 |
+----------+--------------------------------------+
[root@controller ~]# nova list
+--------------------------------------+---------------+--------+------------+-------------+-------------------------------------------+
| ID                                   | Name          | Status | Task State | Power State | Networks                                  |
+--------------------------------------+---------------+--------+------------+-------------+-------------------------------------------+
| 5282ca6c-f232-45e8-8fee-5cd94ab6fcda | cloud_host-01 | ACTIVE | -          | Running     | in-net=192.168.100.101, 202.199.184.102   |
| 6e9402bf-9f1c-402b-bd6c-8ec711af8c5d | cloud_host-10 | ACTIVE | -          | Running     | in-net=192.168.100.103                    |
| 27a462e8-60ef-4d1b-87dd-5a63a22fb300 | cloud_host-20 | ACTIVE | -          | Running     | out-net=202.199.184.105                   |
+--------------------------------------+---------------+--------+------------+-------------+-------------------------------------------+
[root@controller ~]# cinder list
+--------------------------------------+--------+-----------------+------+-------------+----------+--------------------------------------+
| ID                                   | Status | Name            | Size | Volume Type | Bootable | Attached to                          |
+--------------------------------------+--------+-----------------+------+-------------+----------+--------------------------------------+
| 304db00f-3a89-491d-8708-71545606d708 | in-use | cloud_volume-01 | 1    | -           | false    | 5282ca6c-f232-45e8-8fee-5cd94ab6fcda |
| 712a7289-e887-47e3-b6ec-f12971602887 | in-use | cloud_volume-02 | 3    | -           | true     | 6e9402bf-9f1c-402b-bd6c-8ec711af8c5d |
+--------------------------------------+--------+-----------------+------+-------------+----------+--------------------------------------+
[root@controller ~]#
```

图 7.87 云硬盘与云主机连接

访问云主机 cloud_host-01，查看云硬盘连接情况，这里使用了 lsblk 命令，如图 7.88 所示，可以看出云主机 cloud_host-01 多出一块磁盘 vdb，说明云硬盘与云主机已经连接成功。

图 7.88 在云主机上查看云硬盘连接情况

（8）分离云硬盘与云主机

对于云主机不用的磁盘或需要将连接的云硬盘连接到其他云主机上使用时，需要对云硬盘与云主机进行分离。使用帮助命令查看云主机分离云硬盘所使用的命令的具体参数，执行如下命令。

```
[root@controller ~]# nova help volume-detach
usage: nova volume-detach <server> <volume>
Detach a volume from a server.
Positional arguments:
  <server>   Name or ID of server.
  <volume>   ID of the volume to detach.
[root@controller ~]#
```

可以看出使用 nova volume-detach 命令时，分离云主机可以用云主机名称或 ID，但连接云硬盘必须用云硬盘的 ID，执行如下命令。

```
# nova volume-detach cloud_host-01 304db00f-3a89-491d-8708-71545606d708
```

其命令的结果如图 7.89 所示。

```
[root@controller ~]# nova volume-detach cloud_host-01 304db00f-3a89-491d-8708-71545606d708
[root@controller ~]# cinder list
+--------------------------------------+-----------+----------------+------+-------------+----------+--------------------------------------+
|                  ID                  |   Status  |      Name      | Size | Volume Type | Bootable |              Attached to             |
+--------------------------------------+-----------+----------------+------+-------------+----------+--------------------------------------+
| 304db00f-3a89-491d-8708-71545606d708 | available | cloud_volume-01|  1   |      -      |  false   |                                      |
| 712a7289-e887-47e3-b6ec-f12971602887 | in-use    | cloud_volume-02|  3   |      -      |  true    | 6e9402bf-9f1c-402b-bd6c-8ec711af8c5d |
+--------------------------------------+-----------+----------------+------+-------------+----------+--------------------------------------+
[root@controller ~]#
```

图 7.89 分离云硬盘与云主机

2. 对象存储管理

使用 openstack 命令或 swift 命令可以管理对象存储，使用帮助命令可以查看对象存储的相关命令，执行如下命令。

```
# openstack help | grep container
```

或者

```
# swift help | grep container
```

V7-7 对象存储管理

其命令的结果如下。

```
[root@controller ~]# openstack help |grep container
  acl delete        Delete ACLs for a secret or container as identified by its href.
  acl get           Retrieve ACLs for a secret or container by providing its href.
  acl submit        Submit ACL on a secret or container as identified by its href.
  acl user add      Add ACL users to a secret or container as identified by its href.
  acl user remove   Remove ACL users from a secret or container as identified by its href.
  container create  Create new container
  container delete  Delete container
  container list    List containers
  container save    Save container contents locally
  container set     Set container properties
  container show    Display container details
  container unset   Unset container properties
  object create     Upload object to container
  object delete     Delete object from container
  secret container create   Store a container in Barbican.
  secret container delete   Delete a container by providing its href.
  secret container get      Retrieve a container by providing its URI.
  secret container list     List containers.
```

```
[root@controller ~]#
[root@controller ~]# swift help |grep container
no such command: help
    delete              Delete a container or objects within a container.
    download            Download objects from containers.
    list                Lists the containers for the account or the objects
                        for a container.
    post                Updates meta information for the account, container,
                        or object; creates containers if not present.
    stat                Displays information for the account, container,
    upload              Uploads files or directories to the given container.
[root@controller ~]#
```

（1）查看容器及对象列表信息

查看容器及对象列表信息可以使用 openstack 命令，也可以使用 swift 命令，执行如下命令。

```
# openstack container list                //查看容器列表
# openstack object list 容器对象名称       //查看容器对象
```

或者

```
# swift list                              //查看容器列表
# swift list 容器对象名称                  //查看容器对象
```

其命令的结果如图 7.90 所示。

```
[root@controller ~]# openstack container list
+------------------+
| Name             |
+------------------+
| cloud_container-01 |
+------------------+
[root@controller ~]# openstack object list cloud_container-01
+------------------+
| Name             |
+------------------+
| file-01/         |
| file-01/centos命令.txt |
| file-01/img1.jpg |
| file-01/img2.jpg |
| image-01/        |
| image-01/img1.jpg |
+------------------+
[root@controller ~]#
[root@controller ~]#
[root@controller ~]# swift list
cloud_container-01
[root@controller ~]# swift list cloud_container-01
file-01/
file-01/centos命令.txt
file-01/img1.jpg
file-01/img2.jpg
image-01/
image-01/img1.jpg
[root@controller ~]#
```

图 7.90 查看容器及对象列表信息

（2）创建容器

创建容器可以使用 openstack 命令，也可以使用 swift 命令，使用帮助命令可以查看创建容器的相关命令的具体参数，执行如下命令。

```
# openstack help container create
```

或者

```
# swift help post
```

其命令的结果如下。

```
[root@controller ~]# openstack help container create
usage: openstack container create [-h]
```

```
                         [-f {csv,html,json,json,table,value,yaml,yaml}]
                         [-c COLUMN] [--max-width <integer>]
                         [--noindent]
                         [--quote {all,minimal,none,nonnumeric}]
                         <container-name> [<container-name> ...]
Create new container
positional arguments:
  <container-name>       New container name(s)
[root@controller ~]#
[root@controller ~]# swift help post
Usage: swift [--version] [--help] [--os-help] [--snet] [--verbose]
             [--debug] [--info] [--quiet] [--auth <auth_url>]
             [--auth-version <auth_version> |
                --os-identity-api-version <auth_version> ]
             [--user <username>]
             [--key <api_key>] [--retries <num_retries>]
             [--os-username <auth-user-name>] [--os-password <auth-password>]
             [--os-user-id <auth-user-id>]
             [--os-user-domain-id <auth-user-domain-id>]
             [--os-user-domain-name <auth-user-domain-name>]
             [--os-tenant-id <auth-tenant-id>]
             [--os-tenant-name <auth-tenant-name>]
             [--os-project-id <auth-project-id>]
             [--os-project-name <auth-project-name>]
             [--os-project-domain-id <auth-project-domain-id>]
             [--os-project-domain-name <auth-project-domain-name>]
             [--os-auth-url <auth-url>] [--os-auth-token <auth-token>]
             [--os-storage-url <storage-url>] [--os-region-name <region-name>]
             [--os-service-type <service-type>]
             [--os-endpoint-type <endpoint-type>]
             [--os-cacert <ca-certificate>] [--insecure]
             [--no-ssl-compression]
             <subcommand> [--help] [<subcommand options>]

Command-line interface to the OpenStack Swift API.
Positional arguments:
  <subcommand>
    delete              Delete a container or objects within a container.
    download            Download objects from containers.
    list                Lists the containers for the account or the objects
                        for a container.
    post                Updates meta information for the account, container,
                        or object; creates containers if not present.
    stat                Displays information for the account, container,
                        or object.
    upload              Uploads files or directories to the given container.
    capabilities        List cluster capabilities.
    tempurl             Create a temporary URL.
    auth                Display auth related environment variables.
```

[root@controller ~]#

例如，使用 openstack 命令创建容器，执行命令的格式如下。

openstack　container　create　容器名称

其命令的结果如图 7.91 所示。

```
[root@controller ~]# openstack container create cloud_container-10
+-----------------------------------+------------------+--------------------------------------+
| account                           | container        | x-trans-id                           |
+-----------------------------------+------------------+--------------------------------------+
| AUTH_f9ff39ba9daa4e5a8fee1fc50e2d2b34 | cloud_container-10 | txdacec8ebfedb48b5b7a98-00600fec39 |
+-----------------------------------+------------------+--------------------------------------+
[root@controller ~]# openstack container list
+--------------------+
| Name               |
+--------------------+
| cloud_container-01 |
| cloud_container-10 |
+--------------------+
[root@controller ~]#
```

图 7.91　使用 openstack 命令创建容器 cloud_container-10

又如，使用 swift 命令创建容器，执行命令的格式如下。

swift　post　容器名称

其命令的结果如图 7.92 所示。

```
[root@controller ~]# swift post cloud_container-20
[root@controller ~]# swift list
cloud_container-01
cloud_container-10
cloud_container-20
[root@controller ~]#
```

图 7.92　使用 swift 命令创建容器 cloud_container-20

（3）上传文件到容器

上传文件到容器可以使用 swift 命令，执行命令的格式如下。

swift　upload　容器名称

在本地创建文件目录 testfile-01，在目录下创建 one.jpg、two.txt、three.doc 这 3 个文件，用于测试，其操作结果如下。

```
[root@controller ~]# pwd                            //查看本地当前目录
/root
[root@controller ~]# mkdir   testfile-01            //创建本地目录，用于测试使用
[root@controller ~]# cd   testfile-01
[root@controller testfile-01]# pwd
/root/testfile-01
[root@controller testfile-01]# touch   one.jpg  two.txt  three.doc   //创建 3 个文件
[root@controller testfile-01]# ll                   //查看详细信息
total 0
-rw-r--r--. 1 root root 0 Jan 26 05:49 one.jpg
-rw-r--r--. 1 root root 0 Jan 26 05:49 three.doc
-rw-r--r--. 1 root root 0 Jan 26 05:49 two.txt
[root@controller testfile-01]#
```

例如，将测试目录 testfile-01 中的文件 one.jpg 上传至容器 cloud_container-10/testfile-10/ 目录中，执行如下命令。

```
[root@controller testfile-01]# swift   upload   cloud_container-10/testfile-10/   one.jpg
testfile-10/one.jpg
[root@controller testfile-01]#
```

查看容器 cloud_container-10 列表信息，执行如下命令。

```
[root@controller testfile-01]# swift    list    cloud_container-10
testfile-10/one.jpg
[root@controller testfile-01]# openstack    object list    cloud_container-10
+----------------------+
| Name                 |
+----------------------+
| testfile-10/one.jpg  |
+----------------------+
[root@controller testfile-01]#
```

在本地创建目录 testfile-20，将创建的空目录 testfile-20 上传到容器 cloud_container-20 中，执行如下命令。

```
[root@controller testfile-01]# cd ..
[root@controller ~]# mkdir    testfile-20
[root@controller ~]# swift    upload    cloud_container-20    testfile-20
testfile-20
[root@controller ~]#
```

例如，将测试目录 testfile-01 中的文件 two.txt 上传到容器 cloud_container-20 中，将测试目录 testfile-01 中的文件 one.jpg、three.doc 上传到容器 cloud_container-20/testfile-20 目录中，执行如下命令。

```
[root@controller ~]# cd testfile-01
[root@controller testfile-01]# swift    upload    cloud_container-20    two.txt
two.txt
[root@controller testfile-01]# swift    upload    cloud_container-20/testfile-20    one.jpg
testfile-20/one.jpg
[root@controller testfile-01]#swift upload    cloud_container-20/testfile-20    three.doc
testfile-20/three.doc
```

查看容器 cloud_container-20 列表信息，执行如下命令。

```
[root@controller testfile-01]# swift    list    cloud_container-20
testfile-20
testfile-20/one.jpg
testfile-20/three.doc
two.txt
[root@controller testfile-01]# openstack    object list    cloud_container-20
+------------------------+
| Name                   |
+------------------------+
| testfile-20            |
| testfile-20/one.jpg    |
| testfile-20/three.doc  |
| two.txt                |
+------------------------+
[root@controller testfile-01]#
```

在 Dashboard 界面中，先单击"项目"主节点，再单击"对象存储"→"容器"子节点，打开容器列表界面，如图 7.93 所示，可以查看容器 cloud_container-20 的信息。

图 7.93 查看容器 cloud_container-20 的信息

（4）将容器中的文件下载到本地

将容器中的文件下载到本地可以使用 swift 命令，执行命令的格式如下。

swift download 容器名称 文件名目录路径

将容器中的文件下载到本地目录 testfile-01 时，首先要清空目录中的文件。

删除目录 testfile-01 中的所有文件，执行如下命令。

```
[root@controller testfile-01]# rm   one.jpg
rm: remove regular empty file 'one.jpg'? y
[root@controller testfile-01]# rm   two.txt
rm: remove regular empty file 'two.txt'? y
[root@controller testfile-01]# rm   three.doc
rm: remove regular empty file 'three.doc'? y
[root@controller testfile-01]#ll
total 0
[root@controller testfile-01]#
```

将容器 cloud_container-10 中目录 testfile-10 下的 one.jpg 文件下载到本地目录 testfile-01 中，执行如下命令。

```
[root@controller testfile-01]# swift  download  cloud_container-10  testfile-10/one.jpg
testfile-10/one.jpg [auth 1.319s, headers 1.694s, total 1.695s, 0.000 MB/s]
[root@controller testfile-01]# ll
total 0
drwxr-xr-x. 2 root root 20 Jan 26 07:37  testfile-10
[root@controller testfile-01]# cd testfile-10/
[root@controller testfile-10]# ll
total 0
-rw-r--r--. 1 root root 0 Jan 26 05:49 one.jpg
[root@controller testfile-10]#
```

通过以上命令操作结果可以看出，在容器中下载文件时，文件所在的目录也一起下载到本地目录。

（5）删除容器及容器中的文件

删除容器中的文件可以使用 swift 命令，执行命令的格式如下。

swift　delete　容器名称　文件名目录路径

例如，删除容器 cloud_container-10 中 testfile-10 目录下的文件 one.jpg，执行如下命令。

swift　delete　cloud_container-10　testfile-10/one.jpg

其命令的结果如图 7.94 所示。

图 7.94　删除容器 cloud_container-10 中的文件

删除容器时需要保证容器中没有其他文件，否则无法删除容器，可以使用 swift 命令，执行命令的格式如下。

swift　delete　容器名称

例如，删除容器 cloud_container-10，执行如下命令。

swift　delete　cloud_container-10

其命令的结果如图 7.95 所示。

图 7.95　删除容器 cloud_container-10

课后习题

1. 选择题

（1）【多选】目前 Cinder 支持多种后端存储设备，包括（　　）。

　　A. LVM　　　　B. NFS　　　　C. Ceph　　　　D. Sheepdog

（2）Cinder 系统架构中负责调度服务的是（　　）。

　　A. cinder-api　　　　　　　　B. cinder-scheduler

　　C. cinder-volume　　　　　　D. cinder-backup

（3）【多选】Swift 的应用包括（　　）。

　　A. 网盘　　　　　　　　　　B. 备份文档

　　C. IaaS 公有云　　　　　　　D. 移动互联网络和内容分发网络

（4）【多选】Swift 采用的是层次数据模型，存储的对象在逻辑上分为（　　）。

　　A. 账户　　　B. 容器　　　C. 对象　　　D. 服务器

（5）【多选】查看块存储卷列表信息的命令是（　　）。

　　A. openstack　volume　list　　B. cinder　list

C. swift volume list　　　　　　D. cinder volume-list
（6）【多选】查看块存储卷类型表信息的命令是（　　）。
　　A. openstack volume type list　　B. cinder volume-type list
　　C. swift volume-type list　　　　D. cinder type-list
（7）【多选】创建卷类型的命令是（　　）。
　　A. openstack volume type create　B. cinder volume-type create
　　C. cinder type-create　　　　　　D. swift type-create
（8）【多选】创建卷的命令是（　　）。
　　A. openstack volume create　　　B. cinder create
　　C. cinder volume-create　　　　　D. swift volume-create
（9）【多选】删除卷的命令是（　　）。
　　A. openstack volume delete　　　B. cinder volume -delete
　　C. cinder delete　　　　　　　　D. swift volume-delete
（10）【多选】查看容器列表信息的命令是（　　）。
　　A. openstack container list　　　B. swift container –list
　　C. cinder container –list　　　　D. swift list
（11）【多选】创建容器的命令是（　　）。
　　A. openstack container create　　B. swift post
　　C. swift container- create　　　　D. swift create
（12）上传文件到容器中的命令是（　　）。
　　A. openstack container upload　　B. swift upload
　　C. swift container-upload　　　　D. cinder upload
（13）将容器中的文件下载到本地的命令是（　　）。
　　A. openstack container download　B. cinder download
　　C. swift container-download　　　D. swift download
（14）删除容器中的文件的命令是（　　）。
　　A. openstack container delete　　B. cinder delete
　　C. swift delete　　　　　　　　　D. swift container-delete
（15）OpenStack 中的对象存储模块是（　　）。
　　A. Swift　　　B. Cinder　　　C. Nova　　　D. Glance
（16）OpenStack 中的块存储模块是（　　）。
　　A. Swift　　　B. Cinder　　　C. Nova　　　D. Glance

2. 简答题
（1）简述 Cinder 的主要功能。
（2）简述 Cinder 的系统架构。
（3）简述 Cinder 块存储服务与 Nova 计算服务之间的交互。
（4）简述 cinder-api 服务。
（5）简述 cinder-scheduler 服务。
（6）简述 cinder-volume 服务。
（7）简述 cinder-backup 服务。
（8）简述 Cinder 的物理部署。

（9）简述 Swift 的系统架构。
（10）简述 Swift 的应用。
（11）简述存储对象在逻辑上的 3 个层次。
（12）简述 Swift 服务的优势。
（13）简述 Swift 的代理服务器。
（14）简述 Swift 的存储服务器。
（15）简述配置存储策略的步骤。

项目8
OpenStack高级控制服务

【学习目标】

- 掌握Telemetry计量与监控服务基础。
- 掌握Ceilometer数据收集服务。
- 掌握Gnocchi资源索引和计量存储服务。
- 掌握Aodh警告服务。
- 掌握Heat编排服务基础。
- 掌握基于Web界面高级控制服务管理配置方法。
- 掌握基于命令行界面高级控制服务管理配置方法。

8.1 项目陈述

作为一个开源的IaaS平台，OpenStack发展迅速，越来越多的企业基于OpenStack建立了自己的公有云计算管理平台，而计量和监控是必不可少的基础服务。OpenStack起初并不提供这两种服务，需要企业自行开发，为满足这种需求，OpenStack通过Telemetry项目来支持计量服务。Telemetry项目的初衷是支持Telemetry云资源的付费系统，仅涉及付费所需的计量数据，这是由Ceilometer子项目来实现的。除了系统计量外，Telemetry项目也可获取在OpenStack中执行各种操作所触发的事件消息，这是由警告服务Aodh子项目支持的。以前计量数据和事件是一起保存的，现在采用专门的解决方案，即使用资源索引Gnocchi和Panko。Gnocchi提供多项目的时间序列化、度量、资源数据库服务。Panko的目标是提供元数据索引、事件存储服务。

目前OpenStack的性能监控和计量计费成熟度还不高，OpenStack每个版本的相关架构和代码都略有变化。Ceilometer的主要作用是收集OpenStack的性能数据和事件，这对OpenStack的运维非常重要。Gnocchi是与Ceilometer配套的存储系统，Aodh负责监控警告。

OpenStack本身提供了命令行和Horizon供用户管理资源，用户也可编写程序通过REST API来管理云资源，这些方式适合简单、少量资源的管理和单一任务。对于大量资源的管理和复杂的云部署任务，需要使用编排服务（Orchestration）来提高效率。OpenStack的编排服务的项目代号为Heat。Heat是一个通过OpenStack原生REST API基于模板来编排复合云应用的服务。模板的使用简化了复杂的基础设施、服务和应用的定义及部署。模板支持丰富的资源类型，覆盖了Ceilometer的警报、Sahara的集群和Trove的实例等高级资源。它可以基于模板来实现云环境中资源的初始化、依赖关系处理、部署等基本操作，也可以实现自动收缩、负载均衡等高级特性。Heat

提供了一个云业务流程平台，可以让用户使用模板实现资源的自动化部署，更轻松地配置 OpenStack 云体系。

8.2 必备知识

8.2.1 Telemetry 计量与监控服务基础

在 OpenStack 中，用于提供计量和监控服务的项目称为 Telemetry。Telemetry 致力于可靠地收集云中物理和虚拟资源使用的数据，并将这些数据加以保存以便后续查找和分析，当数据满足定义的条件时还会触发相应的处置措施。

1. Telemetry 服务

Telemetry 服务最早从 OpenStack 的 Havana 版本开始出现，只有一个子项目 Ceilometer。Ceilometer 最初的目标很简单，就是提供一个架构，收集所需要的关于 OpenStack 项目的计量数据来支持对用户收费，通过定价引擎使用单一源将事件转换为可计费项目，被称为 "metering"（意为计算）。随着项目的发展，OpenStack 社区发现很多项目都需要获取多种不同类型且不断增长的测量数据，于是 Ceilometer 项目增加了第二个目标，即成为 OpenStack 系统计量的标准方式。不管数据的来源，也不管数据的用途，所有数据采集都可以按照 Ceilometer 的设计来实现。后来因 Heat 对编排项目的需求，Ceilometer 增加了利用获取的计量数据进行警告的功能。

起初 Ceilometer 各类资源的计量数据存储在 SQL 数据库中，随着时间的推移，云环境中计量数据的种类不断增加，数据量增长所带来的性能开销非常大，为此引入了 Gnocchi 解决计量数据的存储问题。Gnocchi 为 Ceilometer 数据提供了更有效的存储和统计分析，以解决 Ceilometer 在将标准数据库用作计量数据的存储后端时遇到的性能问题。Gnocchi 是一个非 OpenStack 项目。

OpenStack 发布 Mitaka 版本时，Ceilometer 的监控警告功能被独立出来作为一个单独的项目 Aodh，目的是让 Ceilometer 专注于数据收集。

Ceilometer 不仅要收集计量项数据，还要收集事件数据，到 OpenStack 发布 Newton 版本时，其将这两类数据分开处理，另一个项目 Panko 提供事件存储服务。

Panko 作为 Ceilometer 项目的一个组成部分，提供事件存储服务，存储和查询由 Ceilometer 产生的事件数据。

Panko 的目标是提供元数据索引、事件存储服务，让用户能够获取特定时间的 OpenStack 资源的状况信息。它为审计和系统调度等用途提供短期和长期数据的可伸缩存储方式。Panko 还包括一种机制，让开发者和系统管理员能够获取关于 Panko 运行状态的报告。

2. Telemetry 服务的逻辑架构

目前，Telemetry 服务被分为 4 个子项目或组件：Ceilometer 负责采集计量数据并进行加工处理，其数据信息主要包括使用对象、使用者、使用时间和用量；Gnocchi 主要用来提供资源索引和存储时间序列计量数据；Aodh 主要提供预警和计量通知服务；Panko 主要提供事件存储服务。Telemetry 服务的逻辑架构如图 8.1 所示。

V8-1 Telemetry 服务

V8-2 Telemetry 服务的逻辑架构

由 Ceilometer 收集和规范化的数据可以发送到不同目标。Gnocchi 用于按时间序列格式获取测量数据、优化存储和查询；Gnocchi 要替换现有的计量数据库接口。另外，Aodh 作为警告服务，用于当有行为违反用户定义规则时发出警告。Panko 作为事件存储服务，用来获取像日志和系统事件行为这样的面向文件的数据。

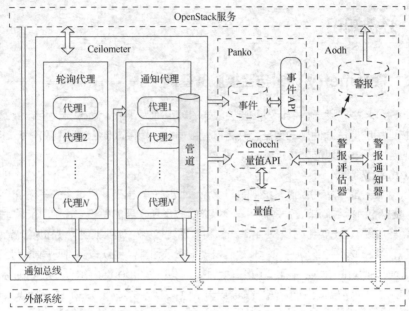

图 8.1　Telemetry 服务的逻辑架构

8.2.2　Ceilometer 数据收集服务

计量对于云计算管理平台的运维是至关重要的，公有云的计费也基于计量数据。Ceilometer 项目专注于数据收集服务，能够对当前所有 OpenStack 核心组件的数据进行规范化处理及传输。Ceilometer 目前正致力于支持未来的 OpenStack 组件。

1．Ceilometer 的主要功能

作为 Telemetry 的组件，Ceilometer 收集的数据为所有的 OpenStack 核心组件提供了用户计费、资源跟踪和警告服务。

Ceilometer 的主要功能如下。

（1）有效轮询 OpenStack 服务相关的计量数据。

（2）通过监测发自服务的通知来收集事件和计量数据。

（3）将收集到的数据发布到多个目标，包括数据存储和消息队列。

2．数据类型计量项和事件

首先了解 Ceilometer 收集的两大类数据，即计量项和事件。

（1）计量项

所谓计量项就是要测量的具体资源属性或项目，又称度量指标。例如，CPU 运行时间就是一个计量项，磁盘读取字节数也是一个计量项。

样值就是采样数据，是某资源时刻某计量项的值。这表示一个计量项的一个可随时间而变化的数值或数据点。例如，CPU 运行时间在某一时刻的值就是一个样值。样值是一次性的，单个样值价值较小，丢失一个样值影响不大，而且样值变化可能非常快。某区间样值的聚合值称为统计值，该值满足给定条件后会产生警告。计量项的值除了被称为样值外，还被称为测量值或计量值。

Telemetry 将计算项分为以下 3 种类型。

① 累计值（Cumulative）：随时间不断增加，如实例使用时数。

② 变化值（Delta）：随时间改变，如网络带宽。

③ 离散值（Gauge）：离散值（如浮动 IP、镜像上载）或者波动值（如磁盘 I/O）。

OpenStack 核心服务都提供自己的计量项集，这些计量项的值可以由 Telemetry 服务轮询获取，也可以直接由 OpenStack 服务发出的通知获取。不同的 OpenStack 版本，其服务支持的计量项也不同，一般新版本会支持更多的计量项。

（2）事件

事件表示 OpenStack 服务中的一个对象在某一时刻的状态，主要包括非数值数据，如一个实例的实例模型或网络地址。一般情况下，事件让用户知道 OpenStack 中一个对象发生了什么改变，如重置虚拟机实例大小、创建一个镜像等。事件值大，信息量多，应当持续收集处理，不能丢失一个事件。

由 Telemetry 服务抓取的事件包括以下 5 个关键属性。

① 事件类型（event_type）：形式为由圆点分隔的字符串，如 compute.instance.resize.start。

② 消息 ID（message_id）：该事件的通用唯一识别码（Universally Unique Identifier，UUID）。

③ 发生时间（generated）：系统中事件发生时间的时间截。

④ 特征（traits）：描述事件的键值对的平面映射。事件的特征包括该事件的大多数细节。特征可以是字符串、整数、浮点数或日期时间类型。

⑤ 原始数据（raw）：主要是为了审计、存储完整的事件消息以便将来评估。

3. Ceilometer 的逻辑架构

Ceilometer 的每项服务都设计成可以水平扩展，Workers 和节点可以根据所需要的负载增加。它通过两个核心的守护进程提供两种数据收集方法。一种是主动发起轮询，由轮询代理守护进程轮询 OpenStack 服务并创建计量项；另一种是被动监听消息队列，由通知代理守护进程监听消息队列上的通知，将它们转换为事件和样值，并应用管道设置的操作。这两个守护进程是由相应的两个核心组件轮询代理和通知代理实现的。Ceilometer 使用基于代理的逻辑架构，收集、规划和重定向数据，用于计量和监测，这些代理之间通过 OpenStack 消息总线进行通信，如图 8.2 所示。

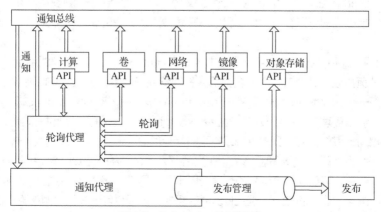

图 8.2 Ceilometer 的逻辑架构

> **注意** 在以前的OpenStack版本中，Ceilometer项目还提供收集器和API服务器组件作为存储和API解决方案。收集器负责接收数据的持久化存储，这个功能可由通知代理替代；API服务器提供数据存储的数据访问。从OpenStack的Newton版本开始，官方不再建议使用Ceilometer，而是推荐使用Gnocchi，以更有效地完成数据的存储和统计分析工作。

（1）轮询代理：请求数据

Ceilometer轮询代理主动向OpenStack服务请求数据，需要获取的是计量项数据，例如，虚拟机实例的CPU的运行时间、CPU的使用率等。轮询代理定期调用一些API或其他工具来收集OpenStack服务的计量信息。这种代理可以配置为轮询本地的虚拟管理器Hypervisor或远程API。轮询频率可通过轮询配置来控制。轮询代理将产生的样值传递到通知代理进行处理。轮询代理如图8.3所示。

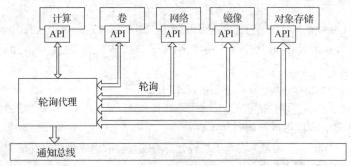

图8.3 轮询代理

这种轮询机制的消耗比较大，可能对API服务影响较大，因而仅用于需优化的端点。另外，这也会带来浪费，如收集的数据可能含有大量重复无用的信息。

Ceilometer轮询代理通过使用在不同命名空间中注册的轮询插件来获取不同种类的计量数据，为不同命名空间提供单一的轮询接口，针对任何命名空间提供轮询支持，目前支持计算代理、中心代理和智能平台管理接口（Intelligent Platform Management Interface，IPMI）代理，只是它们从不同命名空间加载不同的轮询插件来收集数据。可以配置轮询代理守护进程运行一个或多个轮询插件，可使用ceilometer.poll.compute、ceilometer.poll.central和ceilometer.poll.ipmi命名空间的任意组合。单一代理可以在一体化部署中承担两种角色。反之，可以部署一个代理的多个实例以分担负载。

① 计算代理：在每个计算节点中部署，主要通过与Hypervisor的接口调用定期获取资源的使用状态。该代理负责收集计算节点上的虚拟机实例的资源使用数据。这种机制要求与Hypervisor密切交互，因而一个独立的代理类型可完成相关计量项的收集，它部署在物理主机上，可以在本地检索这些信息。计算代理使用计算节点上安装的Hypervisor的API，因此每个虚拟化后端所支持的计量项都不相同，每个检查工具提供一组不同的计量项。

② 中心代理：运行在中心管理服务上，即控制节点上，主要通过OpenStack API获取非计算资源的使用统计信息。可启动多个这样的代理，以水平扩展它的服务。该代理负责轮询公共REST API来检查未能由通知提供的OpenStack资源的额外信息，也可通过简单网络管理协议（Simple Network Management Protocol，SNMP）轮询硬件资源。该代理可以轮询OpenStack网络、

OpenStack 对象存储、OpenStack 块存储和通过 SNMP 的硬件资源。

③ IPMI 代理：IPMI 是管理基于 Intel 结构的企业系统中所使用的外围设备采用的一种工业标准。IPMI 代理负责收集计算节点上的传感器数据和 Intel 节点管理器数据。它要求安装的 ipmitool 工具与 IPMI 节点兼容，通常用于对多种 Linux 发行版的 IPMI 控制。IPMI 代理实例可以安装在每个支持 IPMI 的计算节点上，节点由裸机服务管理且在裸机服务中启用 conductor.send_sensor_data 选项。在不支持 IPMI 传感器或 Intel 节点管理器的计算节点上安装此代理并无大碍，因为该代理检查硬件，如果得不到任何数据将返回空数据。不过出于性能考虑，没有必要这样做。注意，不要在同一个计算节点上同时部署 IPMI 代理和裸机服务。

（2）通知代理：侦听数据

Ceilometer 通知代理监控通知的消息队列，被动获取通知总线上产生的消息，并将其转换为 Ceilometer 的样值或事件数据。通知代理依赖于高级消息队列协议（Advanced Message Queuing Protocol，AMQP）服务，使用者来自 OpenStack 服务和内部通信的通知。系统的核心是通知守护进程，它监控消息队列，获取由其他 OpenStack 组件发送的数据，如 Nova、Glance、Cinder、Neutron、Swift、Keystone 和 Heat 以及 Ceilometer 内部通信等，并对这些数据进行规范化，以发布到所配置的目的地址。通知代理应当部署在一个或多个控制节点上。通知代理如图 8.4 所示。

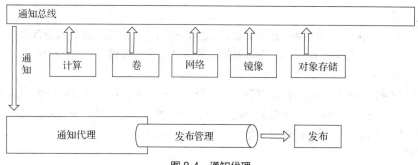

图 8.4 通知代理

所有的 OpenStack 服务发送关于执行的操作或系统状态的通知消息，例如，创建和删除虚拟机实例时会发出对应的通知消息，这些信息是计量或计费的重要依据。有些通知携带可测量的信息，例如，由 OpenStack 计算服务创建的虚拟机实例的 CPU 时间。这是一种被动触发机制，比主动轮询方式的开销小得多。

通知守护进程加载一个或多个侦听器插件，使用命名空间 ceilometer.notification。每个插件几乎都可以侦听任何主题，但是默认侦听 notification.info、notification.sample 和 notification.error。侦听器收集所配置主题的消息，并将其重新发到适当的插件来处理事件和样值。

面向样值的插件列出感兴趣的事件类型，通过一个回调处理相应消息。通过使用通知守护进程的管道，使用回调的注册名称来启用或禁用它。传入的消息被过滤，基于其事件类型，在传递给该回调之前，插件仅能接收感兴趣的事件。

4. 数据处理和管道

在不同场合进行数据测量时，采样要求可能会有所不同。例如，用于计费的数据采样频率会比较低，可能按 10 分钟计；而用于监控的数据采样频率可能很高，达到秒级。不同场合的计量数据发布方式也可能不同，如计费数据要保证完好性和不可否认性，而监控数据就没有这个必要。为解决这些问题，Ceilometer 引入了管道的概念。

（1）管道

管道是一种数据处理机制，在数据源头和相应目标之间转换和发布数据。Ceilometer 中可以同时有多条管道，每条管道都由源头和目标组成。

源是样值和事件数据的生产者，源中会定义测量哪些数据、在哪些端点采集数据以及采样频率等，源实际上是一套通知处理程序。每个源配置封装匹配并映射到一个或多个发布目标的名称。

目标是数据消费者，提供转换发布来自相关源的数据的逻辑。目标定义收到的数据最弱交付给哪些发布器。

Ceilometer 管道由发布器组件组成，Ceilometer 能够获取由代理收集的数据，对它进行处理，然后通过管道以不同的组合发布，这个功能由通知代理实现，如图 8.5 所示。

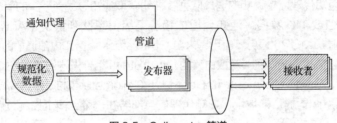

图 8.5　Ceilometer 管道

（2）发布器

Telemetry 服务提供几种传输方法将收集的数据传送到外部系统。数据消费者差异大，就像监控系统一样，数据丢失是可以接受的，而计费系统要求数据可靠传输。Telemetry 提供相应的方法来满足不同类型系统的要求。

发布器组件通过消息总线将数据保存到永久性存储，或者发送给一个或多个外部消费者。一个链可以包含多个发布器。可以为每个数据点配置多个发布器，让同一量值或事件多次发布到多个目的地址，每个发布器可能使用不同的传输系统，如图 8.6 所示。

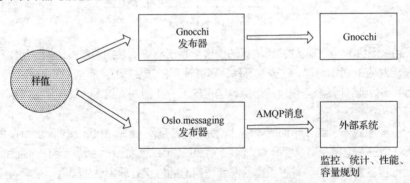

图 8.6　一个样值数据发布到多个目的地址

8.2.3　Gnocchi 资源索引和计量存储服务

Gnocchi 的目标是提供时间序列资源索引和计量存储服务，让用户获取与自己关联的 OpenStack 资源和计量数据。使用由用户定义的存档策略所设置的流动聚合，Gnocchi 提供短期和长期数据的可伸缩存储方式，并基于 Ceilometer 收集的数据提供统计视图。Gnocchi 项目于 2014 年开始推出，目的是作为 Ceilometer 项目的分支，解决 Ceilometer 在将标准数据库用作计量数据的存储后端所遇到的性能问题。

1. Gnocchi 概述

Gnocchi 是 OpenStack 项目的一部分，支持 OpenStack，但也能独立运行。Gnocchi 是开源的时间序列数据库，用于大规模的时间序列数据及资源的存储和索引。这对于云计算管理平台尤其有用，因为云计算管理平台是多项目的，不仅体量大，而且是动态变化的。Gnocchi 的设计充分考虑了这些特点。

Gnocchi 对时间序列存储采取独特的方法，不是存储原始的数据点，而是在存储之前进行聚合计算，这样获取数据非常快，因为只需读取预先计算的结果。这种内置的特性不同于大多数其他时间序列数据库，它们通常将这种机制作为可选的，在查询时计算聚合。Gnocchi 还具备高性能、可扩展和可容错等特性。Gnocchi 对外提供 HTTP REST 接口来创建和操作数据，向操作者和用户提供对计量数据和资源信息的访问渠道。Gnocchi 专门用于存储时间序列及其相关联的资源元数据。因此，它主要的应用场合有计费系统的存储、警告触发、监控系统和数据的统计使用。

2. Gnocchi 的逻辑架构

Gnocchi 的组件包括 HTTP REST API、Metricd 服务和可选的 Statsd 服务，对应的守护进程分别是 gnocchi-api、gnocchi-metricd 和 gnocchi-statsd。Gnocchi 通过 HTTP REST API 或 Statsd 服务接收数据。Metricd 服务在后台对接收到的数据进行统计计算、计量项清理等操作。Gnocchi 的逻辑架构如图 8.7 所示。

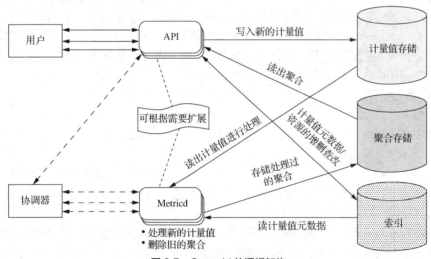

图 8.7　Gnocchi 的逻辑架构

Gnocchi 的所有服务都是无状态的，并且是可以水平扩展的。与许多时间序列数据库相反，可以运行的 Metricd 服务或 API 端点的数量没有限制。如果负载开始增加，只需运行更多的服务（如守护进程）来处理新的请求流即可。如果遇到高可用性场合，也可以采用同样的办法，只需在独立的服务器上启动更多的 Gnocchi 进程。Gnocchi 正常运行还需要 3 个外部组件，即计量存储（Measure Storage）、聚合存储（Aggregated Storage）和索引（Index）。这 3 个外部组件由驱动提供，Gnocchi 以插件方式工作，为这些服务提供不同的选择。

3. Gnocchi 的后端存储

Gnocchi 使用 3 种不同的后端存储数据：用于存储新的即将到来的计量的传入驱动、用于存储时间序列的存储驱动和用于检索数据的索引器驱动。它们分别用于提供计量存储、聚合存储和索引。

（1）传入驱动和存储驱动

传入驱动负责存储传入的计量项的新的计量值。存储驱动负责存储创建的计量项的计量值，接收时间戳和计量值，并且根据定义的归档策略预先计算聚合值。传入驱动默认与存储驱动采用同一个驱动。Gnocchi 可以利用不同的后端存储系统解决传入计量和聚合计量值的存储。根据体系结构的规模，使用文件驱动并将数据存储在磁盘上可能就已足够。如果需要使用文件驱动程序扩展服务器数量，则可以通过网络文件系统（Network File System，NFS）在所有 Gnocchi 进程中导出和共享数据。

（2）索引器驱动

索引器驱动负责存储所有资源的索引、归档策略和度量指标，以及它们的定义、类型和属性。

Gnocchi 要处理的资源和计量值的索引也需要数据库存储，目前支持 PostgreSQL 和 MySQL 这两种数据库驱动。PostgreSQL 具有更高的性能和一些额外的特性，如资源持续时长计算，因而通常是首选的数据库驱动。

8.2.4 Aodh 警告服务

Aodh 的目标是针对 Ceilometer 收集的数据根据定义的规则触发警告。Ceilometer 是从 Ceilometer 中独立出来的 Telemetry 子项目，仅负责警告服务。Aodh 也支持基于 Gnocchi 统计数据的警告设置。Aodh 向上层应用提供 REST API，让上层应用可以通过这些 API 来创建警告策略，以便对云环境中的资源进行实时监控。周期性警告评估保证了监控的颗粒度，警告评估的结果必须是"数据不足""警告""正常" 3 种状态中的一种，策略中还可以定义每种状态触发后采取的下一步动作，例如，监控云服务器，当 CPU 利用率达到 90%时，给指定的统一资源定位系统发送请求，可以发送邮件通知、短信通知等。

1. Aodh 的组件

Aodh 包含以下组件。

（1）API 服务器（aodh-api）：运行于一个或多个中心管理服务器上，提供对存储在数据中心的警告信息的 API 访问接口。

（2）警告评估器（aodh-evaluator）：运行在一个或多个中心管理服务器上，警告评估器周期性地检查警告系统状态，将警告信息通过 RPC 或消息队列 Queue 发送到通知监听器，多个警告评估器需要利用 tooz 协调。

（3）通知监听器（aodh-listener）：运行在一个中心管理服务器上，它侦听事件并根据收到的事件发出警告。针对数据收集服务的通知代理捕获的事件，依据预先定义的规则产生相应的警告。

（4）警告通知器（aodh-notifier）：运行在一个或多个中心管理服务器上，允许根据样值收集的阈值评估来设置警告。它通过 RPC 消息队列接收警告信息，执行相应的操作。

这些服务之间使用 OpenStack 消息总线来通信，共同协作实现警告服务。

2. Aodh 的系统架构

Aodh 的每项服务都设计成可以水平扩展的，可以根据所需的负载扩展，提供守护进程基于定义的警告规则进行评估和通告。

Aodh 的警告组件最早是在 Havana 版本随 Ceilometer 服务一起发布的，到 Liberty 版本时被分离出来作为独立的项目，用来基于样值或事件的阈值评估设置警告。警告可以针对单个量值或多个量值组合进行设置。例如，当一个给定的实例上内存消耗达到 80%时，如果这种情况超过 15 分钟，则触发警告。要设置警告，可以调用 Aodh 的 API 服务器来定义警告的条件和要采取的措施。当然，如果不是云管理员，则只能设置自己组件的警告。警告处置措施有多种方式，但目前只实现以下 3 种方式。

（1）HTTP 回调。当警告被触发时提供一个供调用的 URL。请求的载荷中包括触发警告的原因的详细信息。

（2）日志。将警告存储在日志文件中，这对调试很有用。

（3）Zaqar 消息服务组件。通过 Zaqar API 将通知发送到消息服务。Zaqar 是 OpenStack 内的多项目云消息服务组件，是一个完全的 RESTful API，使用生产者/消费者、发布者/订阅者等模式来传输消息。

3. Aodh 的管理

Aodh 针对 OpenStack 上运行的资源提供面向用户的监控服务。监控类型让用户通过编排服务自动缩小或扩展一组虚拟机实例，也可以将警告用于云资源相关信息。

Aodh 的警告使用的 3 种状态模型含义解释如下。

（1）正常（ok）：管理警告的规则被评估为非触发状态，触发条件不足。

（2）警告（alarm）：管理警告的规则被评估为触发状态，触发条件已满足。

（3）数据不足（insufficient data）：在评估期间没有足够的数据点来判断警告触发状态。

8.2.5 Heat 编排服务基础

作为一个编排引擎，Heat 可以通过基于文本文件形式的模板启动多个复合云应用程序，为 OpenStack 用户提供一种自动创建云应用的方法。Heat 可以兼容亚马逊的 CloudFormation 模板格式，也支持 Heat 自有的 Hot 模板格式。模板的使用简化了复杂的基础设施、服务和应用的定义及部署，许多 CloudFormation 模板可以直接在 OpenStack 环境中运行。

1. 什么是编排服务

所谓的编排，就是按照一定的顺序依次排列。在 OpenStack 环境中，可使用编排服务来集中管理整个云架构、服务和应用的生命周期。编排可以通过预先设定来协调配置同一节点或不同节点的部署资源和部署顺序。

用户将对各种资源的需求写入模板文件，Heat 基于模板文件自动调用相关服务的接口来配置资源，从而实现自动化的云部署。

V8-3　什么是编排服务

在编排服务中，资源特指编排期间创建或修改的对象，可以是网络、路由器、子网、实例、卷、浮动 IP 地址、安全组等。

模板以文本文件的形式描述了云应用的基础设施，主要是需要被创建的资源的细节。Heat 模板支持丰富的资源类型，不仅覆盖了常用的基础架构，如计算、网络、存储、镜像，还覆盖了如 Ceilometer 的警告、Sahara 的集群、Trove 的实例等高级资源。同时，模板可以定义这些资源之间的依赖关系。Heat 读取模板后，自动分析不同资源之间的依赖关系，按照先后顺序依次调用 OpenStack 不同服务或组件的 REST API 来创建资源并部署运行环境，实现相应的业务功能。

Heat 主动调用各 OpenStack 组件的 REST API 创建资源，每次调用模板都会创建一个栈。一个栈往往对应一个应用程序或一个业务项目。在 Heat 项目提供的示例中，WorkPress 就是一个 Web 应用，用它的配置文件可以创建一个栈实例。栈也是云框架中管理一组资源的基本单位，一个栈可以拥有很多资源。

2. Heat 的功能与任务

Heat 的功能与任务如下。

（1）Heat 提供一个基于模板的编排以描述云应用，通过执行适当的 OpenStack API 调用来创建运行的云应用。

（2）Heat 模板定义了资源之间的关系，如某卷连接到某服务器。这使 Heat 能调用 OpenStack API 来创建所有基础设施，按正确的顺序创建全部应用。

（3）Heat 模板以文本文件的形式描述云应用的基础设施，可以由版本控制工具管理。

（4）Heat 与 OpenStack 其他组件整合。模板允许创建大多数 OpenStack 资源类型，如实例、浮动 IP 地址、卷、安全组等，还具有更高级的功能，如实例高可用性、实例动态扩容和嵌套堆栈。

（5）Heat 主要管理基础设施，但是模板也能够与集中配置管理系统 Puppet 和开源运维自动化工具 Ansible 这样的软件配置工具很好地整合。

（6）操作员可以通过安装插件定制 Heat 的功能。

V8-4 Heat 的功能与任务

3. Heat 的工作机制

Heat 的工作机制如图 8.8 所示。

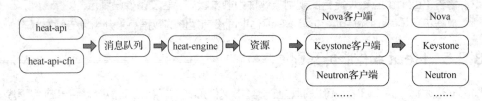

图 8.8 Heat 的工作机制

用户在 Horizon（Dashboard 图形界面）中或者命令行界面中提交包含模板和参数输入的请求，Horizon 或者命令工具会将请求转化为 REST 格式的 API 调用，然后调用 heat-api 或者 heat-api-cfn。heat-api 和 heat-api-cfn 会验证模板的有效性，然后通过 AMQP 异步传递给 heat-engine 来处理请求。

核心组件 heat-engine 提供 Heat 最主要的协作功能，当 heat-engine 收到请求后，会将请求解析为各种类型的资源，每种资源都对应 OpenStack 其他服务的客户端，然后发送 REST 请求给其他服务。通过解析和协作，heat-engine 最终完成请求的处理。

组件 heat-engine 的作用可分为 3 个层面。第一个层面处理 Heat 层面的请求，就是根据模板和输入参数创建栈，这里的栈由各种资源组合而成。第二个层面解析栈中各种资源的依赖关系，以及栈和嵌套的关系。第三个层面就是解析出来的次序、依赖关系和嵌套关系，依次调用各种服务客户端来创建各种资源。

4. Heat 的编排模板

命令行和 Horizon 方式的工作效率并不高，即使把命令行保存为脚本，在 I/O 依赖关系之间仍需要编写额外的脚本来进行维护，而且不易扩展。如果用户直接通过 REST API 编写程序，则同样会引发额外的复杂性。因此，这两种方式都不利于用户通过 OpenStack 进行批量资源管理和编排各种资源。Heat 在这种情况下应运而生，它采用了业界流行的模板方式进行设计和定义编排。用户只需要打开文本编辑器，编写一段基于键值对的模板，就能够方便地得到想要的编排。为了方便用户的使用，Heat 提供了大量的模板例子，通常用户需要选择想要的编排，通过复制、粘贴就能够完成模板的编写。

Heat 的编排方式如下。首先，OpenStack 自身提供基础架构资源，包括计算、网络和存储等。通过编排这些资源，用户可以得到基本的虚拟机。其次，在编排虚拟机的过程中，用户可以编写简单的脚本，以便对虚拟机做简单的配置。再次，用户可以通过 Heat 提供的 Software Configuration 和 Software Deployment 等对虚拟机进行复杂的配置，如安装软件和配置软件等。最后，当用户有一些高级的功能需求，如需要一组能够根据负荷自动伸缩的虚拟机组，或者一组负载均衡的虚拟机时，Heat 提供 Auto Scaling 和 Load Balance 等模板进行支持。

Heat 采用了模板方式来设计和定义编排，用户只需使用文本编辑器编写包含若干节、键值对代码的模板文件，就能够方便地得到所需要的编排。Heat 目前支持两种模板格式，一种是基于 YAML 不是标记语言（YAML Ain't a Markup Language，YAML）格式的 Heat 编排模板（Heat Orchestration Template，HOT）；另一种是基于一种轻量级的数据交换格式 JSON 的模板 CFN （CloudFormation-compatiable），CFN 主要是为了兼容 AWS。HOT 是 Heat 自有的模板格式，资源类型更加丰富，更能体现出 Heat 的特点，比 CFN 更好。

一个典型的 HOT 由以下元素构成。

（1）模板版本：必需项，指定所对应的模板版本，Heat 会根据版本进行检验。

（2）资源列表：必需项，指生成的栈所包含的各种资源。可以定义资源间的依赖关系，例如，生成 Port，并用 Port 来生成虚拟机。

（3）参数列表：可选项，指输入参数列表。

（4）输出列表：可选项，指生成的栈显示出来的信息，可以提供给用户使用，也可以作为输入提供给其他的栈。

V8-5　Heat 的编排模板

8.3　项目实施

8.3.1　基于 Web 界面管理高级控制服务

基于部署远程桌面管理 OpenStack 平台，可以验证和操作高级控制服务管理。用户以云管理员身份登录 Dashboard 界面，可以执行高级控制服务管理操作。

1. 查看项目概况

在 Dashboard 界面中，先单击"项目"主节点，再单击"计算"→"概况"子节点，打开项目概况界面，可以查看项目中资源的使用情况，如图 8.9 所示。

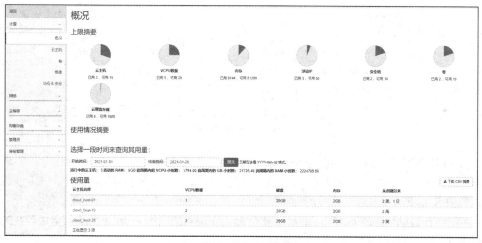

图 8.9　查看项目概况

2. 查看系统概况

在 Dashboard 界面中，先单击"管理员"主节点，再单击"系统"→"概况"子节点，打开系统概况界面，如图 8.10 所示，可以查看系统中资源的使用情况。

图 8.10 查看系统概况

3. 查看资源使用量

在 Dashboard 界面中，先单击"管理员"主节点，再单击"系统"→"资源使用量"子节点，打开资源使用量界面，如图 8.11 所示，可以查看"使用量报表""统计数据"相关情况。

项目	服务	计量	描述	天	平均值	单位
admin	Nova	network.outgoing.packets.rate	虚拟机平均出站流量（packet/秒）	2021-01-26	0.00777034190294	packet/s
admin	Nova	disk.write.requests.rate	写请求的平均速率	2021-01-26	0.0131580089241	request/s
admin	Glance	image	检查镜像是否存在	2021-01-26	1.0	image
admin	Nova	disk.ephemeral.size	临时磁盘大小	2021-01-26	0.0	GB
admin	Nova	disk.read.bytes	读容量	2021-01-26	23665198.5455	B
admin	Nova	network.outgoing.bytes.rate	虚拟机平均出站流量（字节/秒）	2021-01-26	0.417326704777	B/s
admin	Glance	image.size	已上传镜像的大小	2021-01-26	2256971044.57	B
admin	Nova	disk.write.bytes	写容量	2021-01-26	244798.060606	B
admin	Nova	cpu_util	平均CPU利用率	2021-01-26	1.24817325951	%
admin	Swift_meters	storage.objects.containers	容器数目	2021-01-26	1.77966101695	container
demo	Swift_meters	storage.objects.containers	容器数目	2021-01-26	0.0	container
service	Swift_meters	storage.objects.containers	容器数目	2021-01-26	0.0	container
admin	Nova	vcpus	VCPU数量	2021-01-26	1.63636363636	vcpu

图 8.11 查看资源使用量

选择"统计数据"选项卡，可以查看计算（Nova）服务中存在的云主机（instance）、内存容量（memory）、已使用 CPU 时间（cpu）、vCPU 数量（vcpus）等使用情况统计。vCPU 数量统计数据如图 8.12 所示。

选择"统计数据"选项卡，可以查看计算（Nova）服务的网络（Neutron）服务中存在的浮动IP（ip.floating）地址统计数据，如图 8.13 所示。

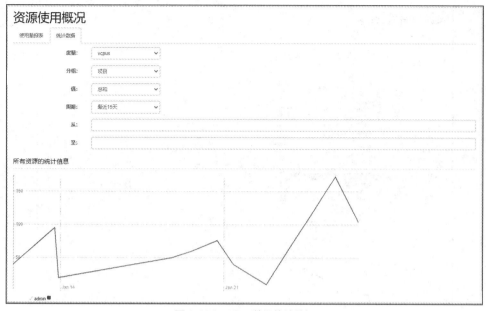

图 8.12 vCPU 数量统计数据

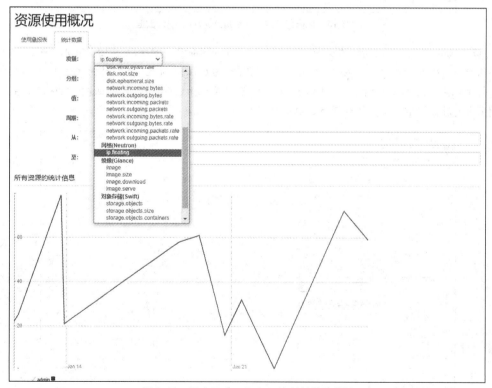

图 8.13 浮动 IP（ip.floating）地址统计数据

选择"统计数据"选项卡，可以检查镜像（Glance）服务中的镜像是否存在（image），查看已上传镜像的大小（image.size）、镜像已下载（image.download）、镜像已使用（image.serve）等使用情况统计。其中，镜像已使用（image.server）统计数据如图 8.14 所示。

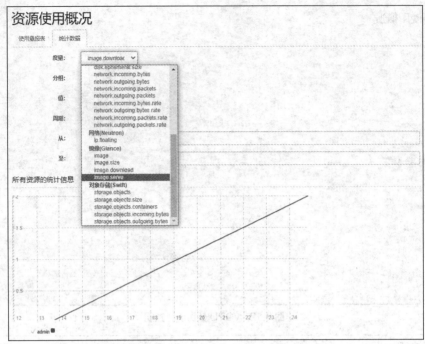

图 8.14 镜像已使用（image.server）统计数据

选择"统计数据"选项卡，可以查看对象存储（Swift）服务中的对象数目（storage.objects）、存储对象总大小（storage.objects.size）、容器数目（storage.objects.containers）、流入字节数目（storage.objects.incoming.bytes）、流出字节数目（storage.objects.outcoming.bytes）等使用情况统计。其中，容器数目统计数据如图 8.15 所示。

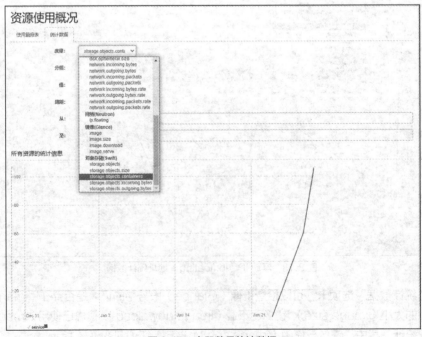

图 8.15 容器数目统计数据

4. 查看系统信息

在 Dashboard 界面中，先单击"管理员"主节点，再单击"系统"→"系统信息"子节点，打开系统信息界面，可以查看系统信息中"服务""计算服务""块存储服务""网络 Agents""Orchestration 服务"等列表相关信息。查看服务列表信息，如图 8.16 所示。

图 8.16 查看服务列表信息

在系统信息界面中，选择"计算服务"选项卡，可以查看计算服务列表信息，如图 8.17 所示。

图 8.17 查看计算服务列表信息

在系统信息界面中，选择"块存储服务"选项卡，可以查看块存储服务列表信息，如图 8.18 所示。

在系统信息界面中，选择"网络 Agents"选项卡，可以查看网络 Agents 列表信息，如图 8.19 所示。

在系统信息页面中，选择"Orchestration 服务"选项卡，可以查看 Orchestration 服务列表信息，如图 8.20 所示。

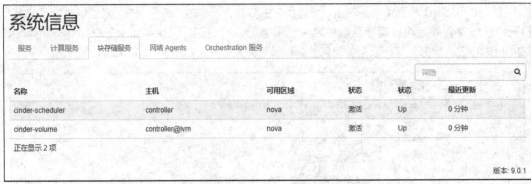

图 8.18　查看块存储服务列表信息

图 8.19　查看网络 Agents 列表信息

图 8.20　查看 Orchestration 服务列表信息

5．云编排

在 Dashboard 界面中，先单击"项目"主节点，再单击"云编排"→"栈"子节点，打开栈界面，可查看栈列表信息，如图 8.21 所示。

图 8.21　查看栈列表信息

在 Dashboard 界面中，先单击"项目"主节点，再单击"云编排"→"资源类型"子节点，打开资源类型界面，可查看资源类型列表信息，如图 8.22 所示。

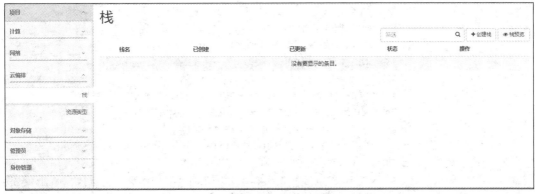

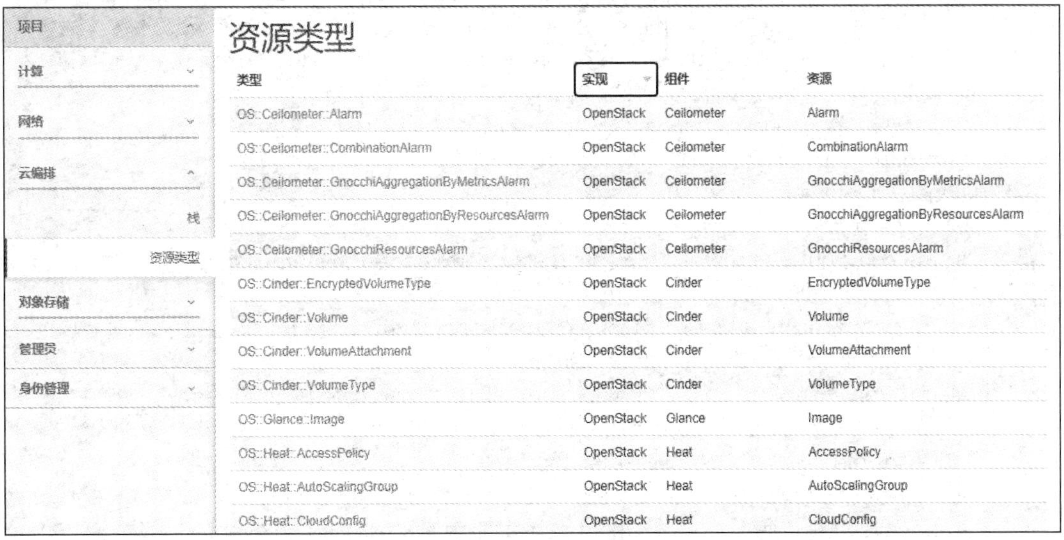

图 8.22　查看资源类型列表信息

8.3.2　基于命令行界面管理高级控制服务

系统管理员一般会基于命令行界面来管理 OpenStack 高级控制服务，可以使用 ceilometer 命令来监控所运行的服务，实时查看平台的运行情况，保障平台运行稳定，维护数据安全，对可能出现的危险做到快速判断和处理。数据主要包括网络数据、实例数据、存储数据和服务资源消耗情况等。可以使用 heat 命令来进行编排服务，包括计算、网络和存储等。

1. 计量监控服务管理

可以使用 ceilometer 命令管理计量监控服务，使用帮助命令可以查看计量监控服务的相关命令，执行命令如下。

```
[root@controller ~]# ceilometer  --help
usage: ceilometer [--version] [-d] [-v] [--timeout TIMEOUT]
                  [--ceilometer-url <CEILOMETER_URL>]
```

V8-6　计量监控服务管理

```
                    [--ceilometer-api-version CEILOMETER_API_VERSION]
                    [--os-tenant-id <tenant-id>]
                    [--os-region-name <region-name>]
                    [--os-auth-token <auth-token>]
                    [--os-service-type <service-type>]
                    [--os-endpoint-type <endpoint-type>] [--os-cacert <cacert>]
                    [--os-insecure <insecure>] [--os-cert-file <cert-file>]
                    [--os-key-file <key-file>] [--os-cert <cert>]
                    [--os-key <key>] [--os-project-name <project-name>]
                    [--os-project-id <project-id>]
                    [--os-project-domain-id <project-domain-id>]
                    [--os-project-domain-name <project-domain-name>]
                    [--os-user-id <user-id>]
                    [--os-user-domain-id <user-domain-id>]
                    [--os-user-domain-name <user-domain-name>]
                    [--os-endpoint <endpoint>] [--os-auth-system <auth-system>]
                    [--os-username <username>] [--os-password <password>]
                    [--os-tenant-name <tenant-name>] [--os-token <token>]
                    [--os-auth-url <auth-url>]
                    <subcommand> ...
Command-line interface to the OpenStack Telemetry API.
Positional arguments:
  <subcommand>
    alarm-combination-create      Create a new alarm based on state of other
                                  alarms.
    alarm-combination-update      Update an existing alarm based on state of
                                  other alarms.
    alarm-create                  Create a new alarm (Deprecated). Use alarm-
                                  threshold-create instead.
    alarm-delete                  Delete an alarm.
    alarm-event-create            Create a new alarm based on events.
    alarm-event-update            Update an existing alarm based on events.
    alarm-gnocchi-aggregation-by-metrics-threshold-create
                                  Create a new alarm based on computed
                                  statistics.
    alarm-gnocchi-aggregation-by-metrics-threshold-update
                                  Update an existing alarm based on computed
                                  statistics.
    alarm-gnocchi-aggregation-by-resources-threshold-create
                                  Create a new alarm based on computed
                                  statistics.
    alarm-gnocchi-aggregation-by-resources-threshold-update
                                  Update an existing alarm based on computed
                                  statistics.
    alarm-gnocchi-resources-threshold-create
                                  Create a new alarm based on computed
                                  statistics.
    alarm-gnocchi-resources-threshold-update
```

	Update an existing alarm based on computed statistics.
alarm-history	Display the change history of an alarm.
alarm-list	List the user's alarms.
alarm-show	Show an alarm.
alarm-state-get	Get the state of an alarm.
alarm-state-set	Set the state of an alarm.
alarm-threshold-create	Create a new alarm based on computed statistics.
alarm-threshold-update	Update an existing alarm based on computed statistics.
alarm-update	Update an existing alarm (Deprecated).
capabilities	Print Ceilometer capabilities.
event-list	List events.
event-show	Show a particular event.
event-type-list	List event types.
meter-list	List the user's meters.
query-alarm-history	Query Alarm History.
query-alarms	Query Alarms.
query-samples	Query samples.
resource-list	List the resources.
resource-show	Show the resource.
sample-create	Create a sample.
sample-create-list	Create a sample list.
sample-list	List the samples (return OldSample objects if -m/--meter is set).
sample-show	Show a sample.
statistics	List the statistics for a meter.
trait-description-list	List trait info for an event type.
trait-list	List all traits with name <trait_name> for Event Type <event_type>.
bash-completion	Prints all of the commands and options to stdout.
help	Display help about this program or one of its subcommands.

See "ceilometer help COMMAND" for help on a specific command.
[root@controller ~]#

（1）查看网络数据

使用帮助命令，查看具体命令使用参数，执行命令如下。

[root@controller ~]# ceilometer help statistics
usage: ceilometer statistics [-q <QUERY>] -m <NAME> [-p <PERIOD>] [-g <FIELD>]
                             [-a <FUNC>[<-<PARAM>]]

List the statistics for a meter.
Optional arguments:
  -q <QUERY>, --query <QUERY>   key[op]data_type::value; list. data_type is
                                 optional, but if supplied must be string,
                                 integer, float, or boolean.

```
  -m <NAME>, --meter <NAME>        Name of meter to list statistics for.
                                   Required.
  -p <PERIOD>, --period <PERIOD>
                                   Period in seconds over which to group samples.
  -g <FIELD>, --groupby <FIELD>
                                   Field for group by.
  -a <FUNC>[<-<PARAM>], --aggregate <FUNC>[<-<PARAM>]
                                   Function for data aggregation. Available
                                   aggregates are: count, cardinality, min, max,
                                   sum, stddev, avg. Defaults to [].
[root@controller ~]#
```

查看网络数据，执行命令如下。

```
[root@controller ~]# ceilometer   statistics   -m   network.incoming.bytes
```

其命令的结果如图 8.23 所示。

```
[root@controller ~]# ceilometer statistics -m network.incoming.bytes
+--------+---------------------------+----------------------------+---------+-----+--------------+-----------+-------
+--------------+---------------------------+----------------------------+
| Period | Period Start              | Period End                 | Max     | Min | Avg          | Sum       | Count
| Duration     | Duration Start            | Duration End               |
+--------+---------------------------+----------------------------+---------+-----+--------------+-----------+-------
+--------------+---------------------------+----------------------------+
| 0      | 2019-12-03T17:50:45.002000 | 2021-01-25T21:37:49.950000 | 41316.0 | 0.0 | 6248.2173913 | 1437090.0 | 230
| 36214564.691 | 2019-12-03T18:01:45.259000 | 2021-01-25T21:37:49.950000 |
+--------+---------------------------+----------------------------+---------+-----+--------------+-----------+-------
+--------------+---------------------------+----------------------------+
[root@controller ~]#
```

图 8.23　查看网络数据

（2）查看实例数据

查看实例数据，执行命令如下。

```
[root@controller ~]# ceilometer   statistics   -m   instance
```

其命令的结果如图 8.24 所示。

```
[root@controller ~]# ceilometer  statistics  -m  instance
+--------+---------------------------+----------------------------+-----+-----+-----+--------+-------
+-------------+---------------------------+----------------------------+
| Period | Period Start              | Period End                 | Max | Min | Avg | Sum    | Count
| Duration    | Duration Start            | Duration End               |
+--------+---------------------------+----------------------------+-----+-----+-----+--------+-------
+-------------+---------------------------+----------------------------+
| 0      | 2019-12-03T17:50:45.002000 | 2021-01-27T13:10:26.067000 | 1.0 | 1.0 | 1.0 | 1391.0 | 1391
| 36356920.83 | 2019-12-03T18:01:45.237000 | 2021-01-27T13:10:26.067000 |
+--------+---------------------------+----------------------------+-----+-----+-----+--------+-------
+-------------+---------------------------+----------------------------+
[root@controller ~]#
```

图 8.24　查看实例数据

（3）查看存储数据

查看存储数据，执行命令如下。

```
[root@controller ~]# ceilometer   statistics   -m   disk.read.requests
```

其命令的结果如图 8.25 所示。

```
[root@controller ~]# ceilometer  statistics  -m  disk.read.requests
+--------+----------------------------+----------------------------+--------+-----+---------------+---
-------+-------+--------------+----------------------------+----------------------------+
| Period | Period Start               | Period End                 | Max    | Min | Avg           | Su
m      | Count | Duration     | Duration Start             | Duration End               |
+--------+----------------------------+----------------------------+--------+-----+---------------+---
-------+-------+--------------+----------------------------+----------------------------+
| 0      | 2019-12-03T17:50:45.002000 | 2021-01-26T22:35:35.740000 | 5850.0 | 0.0 | 1101.56133829 | 29
6320.0 | 269   | 36304430.551 | 2019-12-03T18:01:45.189000 | 2021-01-26T22:35:35.740000 |
+--------+----------------------------+----------------------------+--------+-----+---------------+---
-------+-------+--------------+----------------------------+----------------------------+
[root@controller ~]#
```

图 8.25　查看存储数据

（4）查看服务资源消耗情况数据

查看服务资源消耗情况数据，执行命令如下。

```
[root@controller ~]# ceilometer  statistics  -m  memory
```

其命令的结果如图 8.26 所示。

图 8.26　查看服务资源消耗情况数据

（5）查看资源列表及资源详细信息

查看资源列表，执行命令如下。

```
[root@controller ~]# ceilometer  resource-list
```

其命令的结果如图 8.27 所示。

图 8.27　查看资源列表

查看资源详细信息，执行命令如下。

```
# ceilometer  resource-show  ed75818d-d5c0-4791-87d2-d8bc0fb0a88a
```

其命令的结果如图 8.28 所示。

图 8.28　查看资源详细信息

2. 编排服务管理

可以使用 heat 命令进行编排服务管理，使用帮助命令可以查看编排服务的相关命令，执行命令如下。

V8-7 编排服务管理

```
[root@controller ~]# heat --help
usage: heat [--version] [-d] [-v] [--api-timeout API_TIMEOUT]
            [--os-no-client-auth] [--heat-url HEAT_URL]
            [--heat-api-version HEAT_API_VERSION] [--include-password] [-k]
            [--os-cert OS_CERT] [--cert-file OS_CERT] [--os-key OS_KEY]
            [--key-file OS_KEY] [--os-cacert <ca-certificate-file>]
            [--ca-file OS_CACERT] [--os-username OS_USERNAME]
            [--os-user-id OS_USER_ID] [--os-user-domain-id OS_USER_DOMAIN_ID]
            [--os-user-domain-name OS_USER_DOMAIN_NAME]
            [--os-project-id OS_PROJECT_ID]
            [--os-project-name OS_PROJECT_NAME]
            [--os-project-domain-id OS_PROJECT_DOMAIN_ID]
            [--os-project-domain-name OS_PROJECT_DOMAIN_NAME]
            [--os-password OS_PASSWORD] [--os-tenant-id OS_TENANT_ID]
            [--os-tenant-name OS_TENANT_NAME] [--os-auth-url OS_AUTH_URL]
            [--os-region-name OS_REGION_NAME] [--os-auth-token OS_AUTH_TOKEN]
            [--os-service-type OS_SERVICE_TYPE]
            [--os-endpoint-type OS_ENDPOINT_TYPE] [--profile HMAC_KEY]
            <subcommand> ...

Command-line interface to the Heat API.
Positional arguments:
  <subcommand>
    action-check        Check that stack resources are in expected states.
    action-resume       Resume the stack.
    action-suspend      Suspend the stack.
    build-info          Retrieve build information.
    config-create       Create a software configuration.
    config-delete       Delete the software configuration(s).
    config-list         List software configs.
    config-show         View details of a software configuration.
    deployment-create   Create a software deployment.
    deployment-delete   Delete the software deployment(s).
    deployment-list     List software deployments.
    deployment-metadata-show
                        Get deployment configuration metadata for the
                        specified server.
    deployment-output-show
                        Show a specific deployment output.
    deployment-show     Show the details of a software deployment.
    event               DEPRECATED! Use event-show instead.
    event-list          List events for a stack.
    event-show          Describe the event.
    hook-clear          Clear hooks on a given stack.
```

```
    hook-poll              List resources with pending hook for a stack.
    output-list            Show available outputs.
    output-show            Show a specific stack output.
    resource-list          Show list of resources belonging to a stack.
    resource-mark-unhealthy
                           Set resource's health.
    resource-metadata      List resource metadata.
    resource-show          Describe the resource.
    resource-signal        Send a signal to a resource.
    resource-template      DEPRECATED! Use resource-type-template instead.
    resource-type-list     List the available resource types.
    resource-type-show     Show the resource type.
    resource-type-template
                           Generate a template based on a resource type.
    service-list           List the Heat engines.
    snapshot-delete        Delete a snapshot of a stack.
    snapshot-list          List the snapshots of a stack.
    snapshot-show          Show a snapshot of a stack.
    stack-abandon          Abandon the stack.
    stack-adopt            Adopt a stack.
    stack-cancel-update
                           Cancel currently running update of the stack.
    stack-create           Create the stack.
    stack-delete           Delete the stack(s).
    stack-list             List the user's stacks.
    stack-preview          Preview the stack.
    stack-restore          Restore a snapshot of a stack.
    stack-show             Describe the stack.
    stack-snapshot         Make a snapshot of a stack.
    stack-update           Update the stack.
    template-function-list
                           List the available functions.
    template-show          Get the template for the specified stack.
    template-validate      Validate a template with parameters.
    template-version-list
                           List the available template versions.
    bash-completion        Prints all of the commands and options to stdout.
    help                   Display help about this program or one of its
                           subcommands.
[root@controller ~]#
```

查看编排服务列表，执行命令如下。

```
[root@controller ~]# heat  service-list
```

其命令的结果如图 8.29 所示。

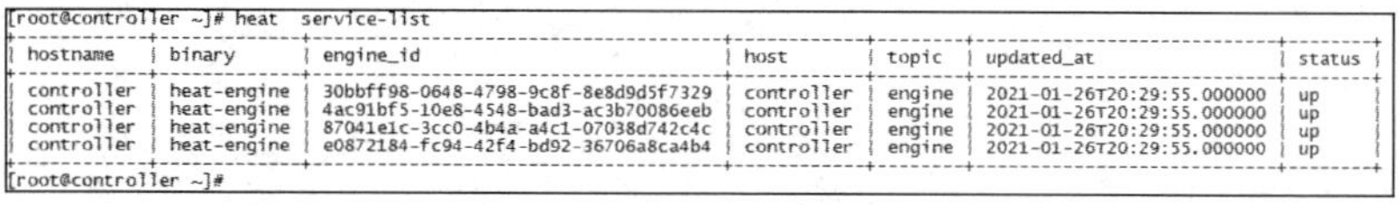

```
[root@controller ~]# heat  service-list
+------------+-------------+--------------------------------------+------------+--------+----------------------------+--------+
| hostname   | binary      | engine_id                            | host       | topic  | updated_at                 | status |
+------------+-------------+--------------------------------------+------------+--------+----------------------------+--------+
| controller | heat-engine | 30bbff98-0648-4798-9c8f-8e8d9d5f7329 | controller | engine | 2021-01-26T20:29:55.000000 | up     |
| controller | heat-engine | 4ac91bf5-10e8-4548-bad3-ac3b70086eeb | controller | engine | 2021-01-26T20:29:55.000000 | up     |
| controller | heat-engine | 87041e1c-3cc0-4b4a-a4c1-07038d742c4c | controller | engine | 2021-01-26T20:29:55.000000 | up     |
| controller | heat-engine | e0872184-fc94-42f4-bd92-36706a8ca4b4 | controller | engine | 2021-01-26T20:29:55.000000 | up     |
+------------+-------------+--------------------------------------+------------+--------+----------------------------+--------+
[root@controller ~]#
```

图 8.29 查看编排服务列表

课后习题

1. 选择题

（1）OpenStack 项目中用来创建栈的组件是（　　）。
　　　A. Nova　　　　B. Cinder　　　　C. Heat　　　　D. Ceilometer
（2）OpenStack 项目中用来计量监控的组件是（　　）。
　　　A. Nova　　　　B. Cinder　　　　C. Heat　　　　D. Ceilometer
（3）【多选】Aodh 的警告使用的状态模型是（　　）。
　　　A. 正常（ok）　　　　　　　　　　B. 警告（alarm）
　　　C. 禁止（forbidden）　　　　　　　D. 数据不足（insufficient data）
（4）【多选】一个典型的 HOT 模板的必需元素是（　　）。
　　　A. 模板版本　　B. 资源列表　　C. 参数列表　　D. 输出列表

2. 简答题

（1）简述 Telemetry 服务包含的子项目，以及各子项目的作用。
（2）简述 Telemetry 服务的逻辑架构。
（3）简述 Ceilometer 的主要功能。
（4）简述 Ceilometer 的逻辑架构。
（5）简述 Gnocchi 的逻辑架构。
（6）简述 Gnocchi 的后端存储。
（7）简述 Aodh 的系统架构。
（8）简述 Aodh 的组件的组成部分。
（9）简述编排服务。
（10）简述 Heat 的目的与任务。
（11）简述 Heat 的工作机制。
（12）简述 Heat 的编排模板。